SOCIETY FOR EXPERIMENTAL BIOLOGY
SEMINAR SERIES: 40

In situ hybridisation: application to developmental biology and medicine

SOCIETY FOR EXPERIMENTAL BIOLOGY SEMINAR SERIES

A series of multi-author volumes developed from seminars held by the Society for Experimental Biology. Each volume serves not only as an introductory review of a specific topic, but also introduces the reader to experimental evidence to support the theories and principles discussed, and points the way to new research.

1. Effects of air pollution on plants. *Edited by T. A. Mansfield*
2. Effects of pollutants on aquatic organisms. *Edited by A. P. M. Lockwood*
3. Analytical and quantitative methods. *Edited by J. A. Meek and H. Y. Elder*
4. Isolation of plant growth substances. *Edited by J. R. Hillman*
5. Aspects of animal movement. *Edited by H. Y. Elder and E. R. Trueman*
6. Neurones without impulses: their significance for vertebrate and invertebrate systems. *Edited by A. Roberts and B. M. H. Bush*
7. Development and specialisation of skeletal muscle. *Edited by D. F. Goldspink*
8. Stomatal physiology. *Edited by P. G. Jarvis and T. A. Mansfield*
9. Brain mechanisms of behaviour in lower vertebrates. *Edited by P. R. Laming*
10. The cell cycle. *Edited by P. C. L. John*
11. Effects of disease on the physiology of the growing plant. *Edited by P. G. Ayres*
12. Biology of the chemotactic response. *Edited by J. M. Lackie and P. C. Williamson*
13. Animal migration. *Edited by D. J. Aidley*
14. Biological timekeeping. *Edited by J. Brady*
15. The nucleolus. *Edited by E. G. Jordan and C. A. Cullis*
16. Gills. *Edited by D. F. Houlihan, J. C. Rankin and T. J. Shuttleworth*
17. Cellular acclimatisation to environmental change. *Edited by A. R. Cossins and P. Sheterline*
18. Plant biotechnology. *Edited by S. H. Mantell and H. Smith*
19. Storage carbohydrates in vascular plants. *Edited by D. H. Lewis*
20. The physiology and biochemistry of plant respiration. *Edited by J. M. Palmer*
21. Chloroplast biogenesis. *Edited by R. J. Ellis*
22. Instrumentation for environmental physiology. *Edited by B. Marshall and F. I. Woodward*
23. The biosynthesis and metabolism of plant hormones. *Edited by A. Crozier and J. R. Hillman*
24. Coordination of motor behaviour. *Edited by B. M. H. Bush and F. Clarac*
25. Cell ageing and cell death. *Edited by I. Davies and D. C. Sigee*
26. The cell division cycle in plants. *Edited by J. A. Bryant and D. Francis*
27. Control of leaf growth. *Edited by N. R. Baker, W. J. Davies and C. Ong*
28. Biochemistry of plant cell walls. *Edited by C. T. Brett and J. R. Hillman*
29. Immunology in plant science. *Edited by T. L. Wang*
30. Root development and function. *Edited by P. J. Gregory, J. V. Lake and D. A. Rose*
31. Plant canopies: their growth, form and function. *Edited by G. Russell, B. Marshall and P. G. Jarvis*
32. Developmental mutants in higher plants. *Edited by H. Thomas and D. Grierson*
33. Neurohormones in invertebrates. *Edited by M. Thorndyke and G. Goldsworthy*
34. Acid toxicity and aquatic animals. *Edited by R. Morris, E. W. Taylor, D.J.A. Brown and J.A. Brown*
35. Division and segregation of organelles. *Edited by S. A. Boffey and D. Lloyd*
36. Biomechanics in evolution. *Edited by J. M. V. Rayner*
37. Techniques in comparative respiratory physiology. An experimental approach. *Edited by C. R. Bridges and P. J. Butler*
38. Herbicides and plant metabolism. *Edited by A. D. Dodge*
39. Plants under stress. *Edited by H. G. Jones, T. J. Flowers and M. B. Jones*

IN SITU HYBRIDISATION: APPLICATION TO DEVELOPMENTAL BIOLOGY AND MEDICINE

Edited by

N. Harris
Department of Biological Sciences, University of Durham, UK

and

D. G. Wilkinson
Laboratory of Eukaryotic Molecular Genetics, National Institute of Medical Research, London, UK

CAMBRIDGE UNIVERSITY PRESS

Cambridge

New York   Port Chester

Melbourne   Sydney

CAMBRIDGE UNIVERSITY PRESS
Cambridge, New York, Melbourne, Madrid, Cape Town,
Singapore, São Paulo, Delhi, Tokyo, Mexico City

Cambridge University Press
The Edinburgh Building, Cambridge CB2 8RU, UK

Published in the United States of America by
Cambridge University Press, New York

www.cambridge.org
Information on this title: www.cambridge.org/9780521380621

© Cambridge University Press 1990

This publication is in copyright. Subject to statutory exception
and to the provisions of relevant collective licensing agreements,
no reproduction of any part may take place without the written
permission of Cambridge University Press.

First published 1990
First paperback edition 2011

A catalogue record for this publication is available from the British Library

Library of Congress Cataloguing in Publication data
In situ hybridisation : application to developmental biology and
   medicine / edited by N. Harris and D. G. Wilkinson.
     p.   cm.—(Seminar series / Society for Experimental Biology :
   40)
   ISBN 0 521 38062 6
   1. In situ hybridisation.   I. Harris, N. (Nicholas)
II. Wilkinson, D. G.   III. Series: Seminar series (Society for
Experimental Biology (Great Britain): 40).
QH452.8.I5   1990
574.87'822—dc20   90-1552   CIP

ISBN 978-0-521-38062-1 Hardback

Cambridge University Press has no responsibility for the persistence or
accuracy of URLs for external or third-party internet websites referred to in
this publication, and does not guarantee that any content on such websites is,
or will remain, accurate or appropriate. Information regarding prices, travel
timetables, and other factual information given in this work is correct at
the time of first printing but Cambridge University Press does not guarantee
the accuracy of such information thereafter.

CONTENTS

List of contributors	*vii*
Preface	*xii*
Non-radioisotopic labels for *in situ* hybridisation histochemistry: a histochemist's view. G. Coulton	1
Use of haptenised nucleic acid probes in fluorescent *in situ* hybridisation A. K. Raap, P. M. Nederlof, R. W. Dirks, J. C. A. G. Weigant and M. van der Ploeg	33
The use of complementary RNA probes for the identification and localisation of peptide messenger RNA in the diffuse neuroendocrine system A. Giaid, S. J. Gibson, J. Steel, P. Facer and J. M. Polak	43
Contributions of the spatial analysis of gene expression to the study of sea urchin development R. C. Angerer, S. D. Reynolds, J. Grimwade, D. L. Hurley, Q. Yang, P. D. Kingsley, M. L. Gagnon, J. Palis and L. M. Angerer	69
Advantages and limitations of *in situ* hybridisation as exemplified by the molecular genetic analysis of *Drosophila* development P. W. Ingham, A. Hidalgo and A. M. Taylor	97
The use of *in situ* hybridisation to study the localisation of maternal mRNAs during *Xenopus* oogenesis H. Perry-O'Keefe, C. R. Kintner, J. Yisraeli and D. A. Melton	115
In situ hybridisation in the analysis of genes with potential roles in mouse embryogenesis D. G. Wilkinson	131
Evolution of algal plastids from eukaryotic endosymbionts G. I. McFadden	143
Localisation of expression of male flower-specific genes from maize by *in situ* hybridisation S. Y. Wright and A. J. Greenland	157
Tissue preparation techniques for *in situ* hybridisation studies of storage-protein gene expression during pea seed development N. Harris, J. Mulcrone and H. Grindley	175

Investigation of gene expression during plant gametogenesis by *in situ* hybridisation
K. G. Jones, S. J. Crossley and H. G. Dickinson   189

Sexing the human conceptus by *in situ* hybridisation
J. D. West   205

Non-isotopic *in situ* hybridisation in human pathology
C. S. Herrington, J. Burns and J. O'D. McGee   241

The demonstration of viral DNA in human tissues by *in situ* DNA hybridisation
M. Wells   271

Index   273

CONTRIBUTORS

Angerer, L. M.
Department of Biology, University of Rochester, Rochester, NY 14627, USA

Angerer, R. C.
Department of Biology, University of Rochester, Rochester, NY 14627, USA

Burns, J.
University of Oxford Nuffield Department of Pathology and Bacteriology, John Radcliffe Hospital, Oxford OX3 9DU, UK

Coulton, G.
Department of Biochemistry, Charing Cross and Westminster Medical School, Fulham Palace Road, London W6 8RF, UK

Crossley, S. J.
School of Plant Science, University of Reading, Whiteknights, Reading RG6 2AS, UK

Dickinson, H. G.
School of Plant Science, University of Reading, Whiteknights, Reading RG6 2AS, UK

Dirks, R. W.
Department of Cytochemistry and Cytometry, Medical Faculty, University of Leiden, Wassenaarseweg, 72, 2333 A1 Leiden, The Netherlands

Facer, P.
Department of Histochemistry, Royal Postgraduate Medical School, Hammersmith Hospital, Du Cane Road, London W12 0NN, UK

Gagnon, M. L.
Department of Biology, University of Rochester, Rochester, NY 14627, USA

Giaid, A.
Department of Histochemistry, Royal Postgraduate Medical School, Hammersmith Hospital, Du Cane Road, London W12 0NN, UK

Gibson, S. J.
Department of Histochemistry, Royal Postgraduate Medical School, Hammersmith Hospital, Du Cane Road, London W12 0NN, UK

Greenland, A. J.
ICI Seeds, Jealott's Hill Research Station, Bracknell, Berkshire RG12 6EY, UK

Grimwade, J.
Department of Biology, University of Rochester, Rochester, NY 14627, USA

Grindley, H.
Department of Biological Sciences, University of Durham Science Laboratories, South Road, Durham DH1 3LE, UK

Harris, N.
Department of Biological Sciences, University of Durham Science Laboratories, South Road, Durham DH1 3LE, UK

Contributors

Herrington, C. S.
University of Oxford Nuffield Department of Pathology and Bacteriology, John Radcliffe Hospital, Oxford OX3 9DU, UK

Hidalgo, A.
Molecular Embryology Laboratory, IRCF, Developmental Biology Unit, Department of Zoology, South Parks Road, Oxford OX1 3PS, UK

Hurley, D. L.
Department of Biology, University of Rochester, Rochester, NY 14627, USA
Present address: Department of Neurobiology and Anatomy, University of Rochester School of Medicine, Rochester, NY 14627, USA

Ingham, P. W.
Molecular Embryology Laboratory, ICRF, Developmental Biology Unit, Department of Zoology, South Parks Road, Oxford OX1 3PS, UK

Jones, K. G.
School of Plant Science, University of Reading, Whiteknights, Reading RG6 2AS, UK

Kingsley, P. D.
Department of Biology, University of Rochester, Rochester, NY 14627, USA

Kintner, C. R.
Molecular Neurobiology, The Salk Institute, 10010 N. Torrey Pines Road, La Jolla, CA 92037, USA

McFadden, G. I.
Plant Cell Biology Research Centre, School of Botany, University of Melbourne, Parkville 3052, Australia

Melton, D. A.
Department of Biochemistry and Molecular Biology, Harvard University, 7 Divinity Avenue, Cambridge, MA 02138, USA

Mulcrone, J.
Department of Biological Sciences, University of Durham Science Laboratories, South Road, Durham DH1 3LE, UK

Nederlof, P. M.
Department of Cytochemistry and Cytometry, Medical Faculty, University of Leiden, Wassenaarseweg, 72, 2333 A1 Leiden, The Netherlands

O'D McGee, J.
University of Oxford Nuffield Department of Pathology and Bacteriology, John Radcliffe Hospital, Oxford OX3 9DU, UK

Perry-O'Keefe, H.
Department of Biochemistry and Molecular Biology, Harvard University, 7 Divinity Avenue, Cambridge, MA 02138, USA

Palis, J.
Department of Biology, University of Rochester, Rochester, NY 14627, USA
Present address: Department of Pediatrics, University of Rochester School of Medicine, Rochester, NY 14627, USA

Polak, J. M.
Department of Histochemistry, Royal Postgraduate Medical School, Hammersmith Hospital, Du Cane Road, London W12 0NN, UK

Contributors

Raap, A. K.
Department of Cytochemistry and Cytometry, Medical Faculty, University of Leiden, Wassenaarseweg, 72, 2333 A1 Leiden, The Netherlands

Reynolds S. D.
Department of Biology, University of Rochester, Rochester, NY 14627, USA

Steel, J.
Department of Histochemistry, Royal Postgraduate Medical School, Hammersmith Hospital, Du Cane Road, London W12 0NN, UK

Taylor, A. M.
Molecular Embryology Laboratory, ICRF, Developmental Biology Unit, Department of Zoology, South Parks Road, Oxford OX1 3PS, UK

van der Ploeg, M.
Department of Cytochemistry and Cytometry, Medical Faculty, University of Leiden, Wassenaarseweg, 72, 2333 A1 Leiden, The Netherlands

Wells, M.
Department of Pathology, University of Leeds, Leeds LS2 9JT, UK

West, J. D.
Department of Obstetrics and Gynaecology, University of Edinburgh Centre for Reproductive Biology, 37 Chalmers Street, Edinburgh EH3 9EW, UK

Wiegant, J. C. A. G.
Department of Cytochemistry and Cytometry, Medical Faculty, University of Leiden, Wassenaarseweg, 72, 2333 A1 Leiden, The Netherlands

Wilkinson, D. G.
Laboratory of Eukaryotic Molecular Genetics, NIMR, The Ridgeway, Mill Hill, London NW7 1AA, UK

Wright, S. J.
ICI Seeds, Jealot's Hill Research Station, Bracknell, Berkshire RG12 6EY, UK

Yang, Q.
Department of Biology, University of Rochester, Rochester, NY 14627, USA

Yisraeli, J.
Department of Biochemistry and Molecular Biology, Harvard University, 7 Divinity Avenue, Cambridge, MA 02138, USA

PREFACE

Advances in our understanding of biological mechanisms have frequently been associated with the development of new techniques, and this is certainly true for the method of *in situ* hybridisation. Although the methods of molecular biology have led to the identification and characterisation of innumerable genes and their mechanisms of regulation, a crucial piece of evidence was lacking in early work: detailed knowledge of the sites of gene expression in tissues. In essence, *in situ* hybridisation combines histochemistry with molecular biology, and enables the rapid analysis of the distribution of RNA (or DNA) in tissues. This information is proving particularly important in the field of developmental biology, since a fundamental aspect of development is the spatial and temporal expression of genes. The application of *in situ* hybridisation is leading to a revolution in our understanding of plant and animal development, in particular when the technique is combined with other approaches, for example, genetics. In addition, the sensitivity and specificity of *in situ* hybridisation has found application in the field of medicine, where it is giving new insights into the functioning of healthy tissues and the diagnosis and study of diseases.

The two-day meeting on '*In situ* hybridisation', held during the SEB Edinburgh Conference in April, 1989, was the first venture to be sponsored jointly by the SEB and the Royal Microscopical Society; it also received support from DuPont, ICI Seeds and Shell. The sessions brought together workers in the fields of development and medicine, and emphasised practical aspects and variations of the technique, and both the strengths and limitations of its applications. All of the speakers have contributed to this volume. Chapters 1 and 2 discuss non-radioactive methods of *in situ* hybridisation, and Chapter 3 details radioisotopic methods and their application to studies of neuropeptides. Chapters 4–7 describe the ways in which *in situ* hybridisation has led to advances in understanding developmental mechanisms in sea urchins, fruit flies, frogs and mice. The plant papers (Chapters 8–11) cover evolutionary aspects of endosymbiosis, studies of floral biology and the incompatibility mechanism, and aspects of seed biology. Finally, Chapters 12–14 describe some recent medical applications in the sexing of embryos and the analysis of viral infections.

We have sought to produce a volume which both reviews progress and will help the reader and would-be exponent through the various options and variations of the technique. To this end we asked authors to include, where appropriate, a protocol section at the end of their chapter.

Finally, we give our thanks to the authors who gathered from Europe, North America and Australia and gave freely of their time, ideas and experience both to the Conference and to this volume.

David G. Wilkinson
Nick Harris
October, 1989

Gary Coulton

Non-radioisotopic labels for *in situ* hybridisation histochemistry: a histochemist's view

Introduction

The early development of *in situ* hybridisation relied heavily upon the use of radioisotopes and autoradiography for the visualisation of specifically hybridised gene probes (Buongiorno-Nardelli & Amaldi, 1969; Gall & Pardue, 1969, 1971). Radioisotopic labelling retains an important rôle, especially where maximum sensitivity is required, but in recent years a group of techniques employing non-radioisotopic labelling technology have been developed. They have been designed to avoid the perceived drawbacks of radiolabels in terms of safety and inconvenience.

In this review I hope to provide a flavour of the range of non-isotopic techniques available, to discuss their relative merits and to divine future developments. Some general methodologies are outlined but these must be considered as general guidelines from which optimal protocols must be individually determined (some detailed protocols are provided in later chapters).

Aim

In situ hybridisation identifies specific DNA or RNA sequences in tissue sections. What questions can *in situ* hybridisation answer that other methods cannot?

Standard molecular biological methods using gel electrophoretic separation of restriction fragments is ideal for characterising genomes, for viral detection or general changes in gene expression. However, in certain respects these methods have severe limitations. Extremely small quantities of nucleic acid can be detected by filter hybridisation but the identity of the cell containing the target sequence is unknown. Therefore electrophoretic separation methods provide no distributional information but are rather an accurate measure of the average nucleic acid content of a tissue homogenate. Only *in situ* hybridisation provides the capacity for the identification of specific target sequences within a mixed cell population. This depends upon the presence of localised areas of high target sequence concentration within a background of non-homologous sequences. Particular examples may be seen in the localisation of tissue-specific mRNA transcripts (Dirks et al., 1989), viral (Burns et al., 1988) or bacterial infection (Näher, Petzoldt & Sethi, 1988), oncogenes in small cell lung carcinoma (Gu et al., 1988), sex determination (Burns et al., 1985) or chromosomal abnormalities (Hopman et al., 1988) and physical gene mapping (Bhatt et al., 1988).

How is it done?

Essentially all hybridisations depend on the fundamental characteristic of nucleic acids; that is single strands of DNA, made up of complementary base sequences, bind together more tightly than non-homologous sequences. All hybridisation methods depend upon a thorough understanding of the physico-chemical forces which control hybridisation.

Types of nucleic acid hybrid:

DNA–DNA	Increasing stability
DNA–RNA	at a particular
RNA–RNA	temperature

Double stranded (DS) DNA, single stranded (SS) DNA, DNA oligonucleotides or RNA may be used as probes for hybridisation.

Factors affecting *in situ* hybridisation sensitivity

The factors affecting hybrid stability and hybridisation kinetics, and thereby sensitivity, have been well established for duplexes in solution, but there are important differences between hybrids in solution and as immobilised duplexes, particularly so in tissue sections. Final signal strength is determined by a number of interdependent factors:

1. Probe type: DNA, RNA or oligonucleotide
2. Specific activity and type of labelling system
3. Inherent sensitivity of detection system
4. Optimum hybridisation conditions
5. Target retention and accessibility
6. Degree of non-specific background inherent to the detection system.

Theoretical background

Complementary DNA strands are bound by relatively weak, non-covalent bonds including hydrogen bonds between guanine and cytosine (GC) and adenine and thymine (AT) base pairs; given sufficient energy these bonds break, denaturing the duplex.

The temperature at which half of a population of DNA duplexes denatures can be described by the value of $T_m(°C)$ (Thomas & Dancis, 1973). This temperature varies depending upon several physical characteristics of the nucleic acid strands involved and is a crucial factor in the determination of optimal hybridisation conditions.

T_m is calculated from the equation:

$$T_m = 81.5°C + 16.61 \log M + 0.41(\%GC) - 820/L - 0.6(\%F) - 1.4\%$$

(Mis), where: M = ionic strength (M/L), %GC = % Guanine/cytosine, L = probe length (bp), %F = %formamide (duplex stabiliser), % Mis = degree of non-complementarity

Increasing ionic strength stabilises duplexes and reduces the T_m of a particular duplex. The proportional GC content affects hybrid stability because GC base pairs are joined by three hydrogen bonds rather than two joining AT pairs. Clearly, long complementary pieces of duplex contain more bonds than short ones and are more stable. RNA–RNA duplexes are 10 to 15°C more stable than DNA–DNA duplexes of similar length and composition (Cox et al., 1984) whilst DNA–RNA duplexes have intermediate T_ms (Casey & Davidson, 1977). As T_m is also dependent on probe length, it follows that lower hybridisation temperatures are needed for oligonucleotide probes. Hydrogen bonds between complementary nucleotides, especially in DNA duplexes, can be disrupted by formamide thus destabilising the duplex and lowering the T_m. Clearly the effect on non-homologous duplexes is greater than on homologous strands. Formamide is added to hybridisation mixtures so that optimal stringencies are possible at temperatures less deleterious to tissue sections.

In practice, in situ hybridisations are carried out at about $T_m - 25°C$. Such low stringency produces maximum probe hybridisation but also allows some hybridisation to non-homologous sequences. These more weakly bound duplexes can be preferentially denatured by low salt post-hybridisation washes at temperatures nearer to T_m. Hybridisation times are often determined empirically; however, knowledge of hybridisation kinetics and the factors which affect them can be used as a preliminary guide.

The time needed for half a population of probe to hybridise to the target sequence may be described in terms of $t_\frac{1}{2}$ which is calculated by the equation:

$$t_\frac{1}{2}(s) = \frac{N \ln 2}{3.5 \times 10^5} \times L^{0.5} \times C$$

where N = sequence complexity, L = probe length, C = [probe] (M/L).

In solution this relationship suggests that most rapid hybridisation occurs with fairly long, simple probes at high concentration. However, problems of probe penetration through tissue mean that shorter probes, of the order of 100 bp give the most rapid hybridisation for in situ hybridisation. In addition, penetration of probes may be further compromised by labels attached to the probe. This is due to steric hindrance and increased rigidity which can alter the shape, charge density and distribution of the probe.

Probe production

A detailed description of the microbiological procedures employed for the production of the various types of probe is outside the terms of reference for this chapter and can be seen in Maniatis, Fritsch & Sambrook (1982). However, an appreciation of the processes involved is important for the end user.

In most instances you will wish to localise previously characterised DNA or RNA sequences. The traffic in gene probes is very free at present but more often than not a donor laboratory will send only a small quantity (μg) of recombinant plasmid DNA or riboprobe vector in ethanol or buffer, or possibly as an agar stab preparation of recombinant *Escherichia coli*. Such quantities of DNA are enough for only a few *in situ* hybridisations and so it is necessary to amplify the amount of probe for future use.

Until very recently the methods of preparation, purification and labelling of sufficient quantities of pure probe have involved time-consuming, expensive and, very often, infuriating microbiological procedures which are beyond the logistic capabilities of many laboratories. Outlined below are some of the various steps involved in DNA production by standard molecular methods, but this is by no means exhaustive.

1. Quality control electrophoresis
2. Bacterial transformation.
3. Small-scale bacterial culture
4. Quality control electrophoresis
5. Large-scale bacterial culture
6. Plasmid purification
7. Restriction digest
8. Quality control electrophoresis
9. Preparatory electrophoresis
10. Band elution
11. Probe labelling (e.g. nick translation).

Even in experienced hands this process can take a couple of weeks.

The Polymerase Chain Reaction (PCR) (Saiki *et al.*, 1986) has altered the picture to the potential benefit of the end user. PCR can, under optimal conditions, produce micrograms of vector-free probe from a nanogram or less of DNA template. Recently PCR was used for direct production of biotinylated DNA probes (Lo, Mehal & Fleming, 1988) though the polymerase reaction efficiency is slightly reduced. Insert amplification depends on the use of specifically synthesised oligonucleotide primers complementary to sequences on alternate strands and at either end of the cloning site of the plasmid vector (Fig. 1). In the presence of an excess of

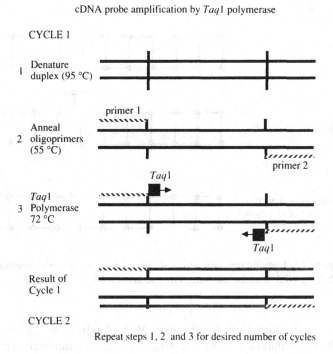

Figure 1. Schematic representation of the Polymerase Chain Reaction (PCR) for the amplification of cDNA inserts within recombinant plasmids.

nucleotides, from the bacterium *Thermus aquaticus* (Taq 1) polymerase will make a copy of the insert in the first cycle. This doubles the insert seed strands for the next polymerase cycle and so on. The geometric increase in insert copies soon produces the required amount of pure probe. Less than 1 ng of template DNA should be used in order to prevent rapid depletion of dNTP substrates. This method is successful for insert sequences of up to 2.5 kb. If PCR fulfils our expectations, it will make the histochemist independent by removing our reliance, however agreeable, upon molecular biologists.

Probe labelling methods

DNA probes

Two main principles underlie most of the methods for probe visualisation. Firstly, labelled nucleotides are enzymatically incorporated into probes during synthesis or afterwards by, for example, nick translation and visualised by standard immunocytochemical protocols. Secondly, probes may be chemically modified and visualised by binding to an immunogenic hapten and indirect antibody staining.

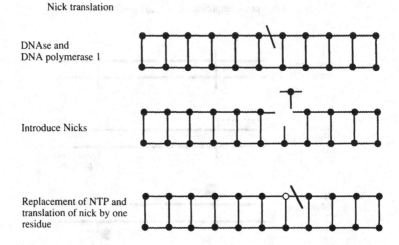

Figure 2. Schematic representation of the reactions involved in labelling of DNA by Nick translation using DNAse and DNA polymerase 1.

Nick translation (double-stranded DNA), Rigby et al., 1977)

DNAse 1 makes cuts in one strand (nicks) at random sites on template DNA and DNA polymerase, via its 5'–3' exonuclease and 5'–3' polymerase activities, fills in the nicks with free deoxynucleotidephosphates (dNTPs) (Fig. 2). Labelled dNTPs are incorporated by this activity and, because the nicks are randomly sited, denaturation of the duplex produces strands of differing lengths. The specific activity of labelled probes depends on the extent of replacement by labelled nucleotides. Labelling can take between 30 minutes and 3 hours but long incubations cause degradation of labelled DNA. Gene-specific insert sequences may be purified and labelled or, alternatively, intact recombinant plasmid may be nick translated. This does not affect specificity and actually increases signal strength due to 'networking' of junction pieces (Lawrence & Singer, 1985).

Advantages
Control over probe size, yield, specific activity, substrate concentration and reaction time
60–70% incorporation of label
Standard hybridisation temperatures can be used
Large yield even of biotinylated probes
Labels circular and linear DNA
Uniform labelling.

Disadvantages
Unpredictable with impure template DNA
Uniform labelling not restricted to insert DNA
Needs large amounts of substrate
Requires careful control of time and temperature
Probe is not strand specific. Strand reannealing of probe may occur
Will not label single-stranded DNA.

Oligonucleotide primer extension synthesis
Random or unique oligonucleotide primers can be used with Klenow polymerase to synthesise new DNA strands complementary to a template strand, starting at the free 3'-hydroxyl terminus of the primer. It is essential to use a polymerase with no 5'-3' exonuclease activity otherwise the primers will be degraded (Feinberg & Vogelstein, 1983, 1984).

Advantages
High specific activity
Efficient utilisation of label
Flexible reaction temperature and time (up to overnight)
Can label small amounts of DNA
Will label single stranded DNA
Incorporated label not excised during reaction
Probe size is uniform and controllable.

Disadvantages
Average yield only about 70 ng (i.e. only one hybridisation)
Relatively inefficient with circular DNA substrates
Labelling not limited to insert sequence.

Photobiotin labelling
Biotin is linked via a spacer molecule to a terminal aryl azide, which when activated to a highly reactive aryl nitrene by strong light, covalently links to the nucleotides in DNA or RNA (Fig. 3) (Forster et al., 1985).

Advantages
Simple protocol
No enzymes involved.

Disadvantages
No real control over rate of incorporation (usually about 1 in every 100 nucleotides).

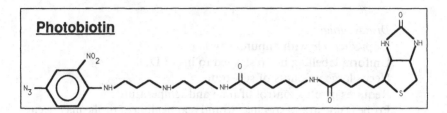

Figure 3. The molecular structure of photobiotin. Biotin is bound to a aryl azide group via a long spacer arm to reduce steric hindrance. Strong light converts the aryl azide to highly reactive aryl nitrene which binds to cytosine moieties in nucleic acids.

Labelling of single stranded DNA

Single-stranded DNA probes can be made *in vitro* using the M13 bacteriophage vectors. Potentially these probes are superior to double-stranded probes because they cannot reanneal. So far, limited use has been made of these probes owing to the difficulty of vector construction and the vagaries of probe synthesis and labelling (Varndell *et al.*, 1984).

Oligonucleotides

Increasingly, oligonucleotides are used for *in situ* hybridisations (for a review see Lewis, Sherman & Watson, 1985). They can be simply and reliably synthesised using standard phosphotriester or phosphoramidite chemistry given the availability of an automatic synthesiser. Oligonucleotides have a number of advantages over cloned DNA probes including:

Higher and controllable specific activities
Amino acid sequences can act as blueprints in the absence of DNA sequence data
Multiple probes can be generated for a single sequence. Thus specificity can be tested by signal strength and cellular distribution
Sense probes can be used as very specific negative controls.

Oligonucleotides may be 5' end-labelled by T4 polynucleotide kinase (Richardson, 1981) or 3' end-labelled by terminal deoxynucleotide transferase activity (Bollum, 1974; Maniatis *et al.*, 1982). High specific activity can be achieved by end-labelling with biotinylated dNTPs. Densely biotinylated oligonucleotides suffer from steric hindrance reducing hybridisation efficiency.

RNA probes

Detailed descriptions of the production and use of riboprobes have been given elsewhere (e.g. Angerer *et al.*, 1987 and in Chapters 3 & 4). Only

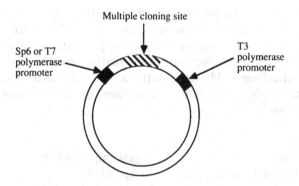

Figure 4. Diagram of the organisation of a typical riboprobe showing the orientation of Sp6 and T7 polymerase promoter sequences.

those features pertinent to the central theme of this essay will be described here. Single-stranded RNA probes can be made *in vitro* from RNA transcription vectors which contain promoter sequences for RNA polymerases such as SP6, T3 and T7 (Cox *et al.*, 1984) (Fig. 4). Antisense riboprobes are far more sensitive with lower background than double-stranded DNA probes because they do not suffer from probe reannealing. Other advantages of riboprobes over cDNAs are that they can be labelled to higher specific activity, they are thermally stable and have constant and defined probe size, all of which favours increased sensitivity and consistency of results. However, so far there are relatively few reports of the use of non-radioactively labelled RNA probes. This may be because biotinylated nucleotides are poor substrates for RNA polymerases (Höfler, 1987). A comparison has been made between biotinylated and [^{35}S] labelled riboprobes complementary to portions of human papilloma virus-16 RNA for use in *in situ* hybridisation (Crum *et al.*, 1988). Biotinylated riboprobes were transcribed in GEM 1 vectors with either T7 or Sp6 polymerase. Sp6 transcription was also used by Zabel & Schäfer (1988) to make biotinylated riboprobes to localise calcitonin and calcitonin-gene related peptide mRNAs in rat parafollicular cells. Neither of these studies reported problems with biotin incorporation but Crum and co-authors indicated lower sensitivity of the biotin labelled riboprobe Allyl-UTP and subsequent biotinylation has been tried with limited success due to high background problems (Höfler, 1987). Photobiotin methods are also being adapted for RNA probes, and some of the chemical modification methods such as sulphonation or mercuration may prove useful in releasing the true potential of this powerful system.

Non-radioactive labelling systems

Despite the power of *in situ* hybridisation using radiolabelled probes there are serious limitations. Many isotopes, for example ^{32}P, have short functional half-lives, are expensive, and require stringent safety controls for workers and disposal. The development of alternative sensitive, safe and rapid label systems has been highly desirable (for a review see van der Ploeg *et al.*, 1986). The main advantages and disadvantages of non-radioisotopic labels are listed below.

Advantages
Probe stability	Signal amplification
Safety (handling and disposal)	Inexpensive
Rapid detection	High resolution (E.M.)
No specialist labs	Quantifiable (in certain cases)
Multiple, simultaneous labelling	Compatible with immunohistochemistry.

Disadvantages
Decreased hybridisation efficiency.	Claimed low sensitivity.

Labelling systems currently used for *in situ* hybridisation include:
Biotin–avidin (streptavidin) conjugates, Biotin–anti-biotin conjugates, Photobiotin–avidin conjugates, Sulphonated nucleotide–antibody conjugates, Mercurated nucleotide–antibody conjugates, Acetylaminofluorenyl–antibody conjugates, Bromodeoxyuridine–antibody conjugates, Fluorochrome-labelled nucleic acids and anti-DNA:RNA hybrid antibodies.

Biotin–avidin labelling systems are the most commonly used non-radioisotopic technique. Localisation of hybridised probe depends upon the high binding constant of biotin (Vitamin H derived from egg white) for avidin (glycoprotein from egg yolk). The association constant ($10^{15} M^{-1}$) is some 10^6 times greater than typical antigen–antibody affinities. Biotin can be enzymatically attached to the C-5 position of pyrimidine rings of nucleic acids via an allylamine linker arm (Langer, Waldrop & Ward, 1981) but high-quality biotinylated nucleotide phosphates can be obtained commercially (Enzo or Sigma) for use in labelling reactions. DNA polymerases will readily use biotinylated bases as substrates but RNA polymerases utilise biotinylated nucleotides less efficiently. Biotin is usually bound to the probe via a long spacer molecule (e.g. Biotin 11-UTP), in order to reduce steric hindrance and increase hybridisation efficiency. Biotinylated probes

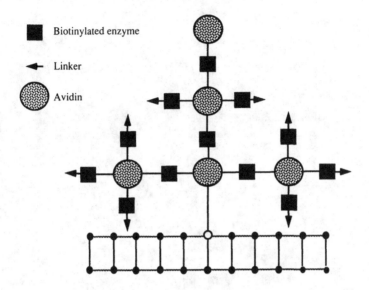

Figure 5. The principle of biotinylated probe visualisation by application of avidin–biotin–polyenzyme conjugates. This combination produces considerable signal amplification from specifically hybridised probes.

detected immunocytochemically by specific anti-biotin antibodies were used initially for gene localisations on *Drosophila* polytene chromosomes (Langer & Ward, 1981; Langer-Safer, Levine & Ward, 1982) and on mammalian metaphase chromosomes (Hutchinson *et al.*, 1982; Mannuelidis, Langer-Safer & Ward, 1982). The site of a specifically hybridised biotinylated probe may also be localised by binding to avidin linked to a variety of signal generating reporter molecules. Since avidin has four binding sites for biotin whilst biotin has only one avidin binding site, non-specific binding to endogenous biotin may be blocked (Fig. 5). Streptavidin is an avidin-like molecule derived from *Streptomyces avidinii* which, because of its neutral isoelectric point, has a very low non-specific binding to DNA. Simple systems, where an avidin-label conjugate is bound to immobilised biotin on the probe, can be amplified by using controlled applications of streptavidin/biotin–alkaline phosphatase complexes. Figure 6 shows localisation of the Y chromosome-specific testis determining factor locus using a biotinylated double stranded genomic DNA probe and visualisation by an avidin–alkaline phosphatase complex (Vector Laboratories). This method relies on the fact that alkaline phosphatase has several biotin binding sites enabling

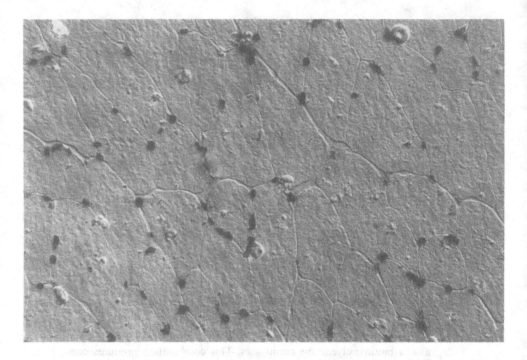

Figure 6. Shows localisation of the testis determining factor locus on the Y chromosome, within the nuclei of mouse skeletal muscle fibres, using a biotinylated double stranded DNA probe. The reporter molecule in this case was an avidin–alkaline phosphatase complex.

the enzyme to interact with more than one streptavidin molecule. Pre-formed complexes with several enzymes are commercially available (Amersham, Enzo).

Bromodeoxyuridine can be immunohistochemically detected when incorporated into DNA and is a good substrate for both eukaryotic and prokaryotic DNA polymerases (Traincard et al., 1983; Boultwood et al., 1986).

DNA probes can be chemically modified by insertion of an antigenic sulphone group in cytosine moieties and visualised by a double-antibody immunohistochemical reaction (Verdlov et al., 1974) (Chemiprobe, FMC BioProducts). Sulphonation of DNA is very quick and does not involve enzymatic manipulation of DNA.

Non-enzymatic introduction of mercury at the C5 position of pyrimidine bases is a highly efficient labelling system for DNA probes (Hopman et al., 1986). After hybridisation the probe is detected by reaction with a sulphydryl–hapten (trinitrophenyl (tnp) ligands). The tnp hapten is in turn detected by a specific antibody-fluorochrome conjugate. A possible alternative might be to use mercurated nucleotides for either nick translation, PCR or riboprobe transcription as they are suitable substrates for DNA and RNA polymerases.

Bauman et al. (1981) directly coupled fluorochromes onto the 3' termini of riboprobes with some success.

Antigenic 2-acetylaminoflurenyl groups can also be incorporated into nucleic acids by treatment with N-acetoxy-2-acetylaminofluorene. Modified probes are localised by indirect immunofluorescence with antibodies against the AAF-modified guanosine (Landegent et al., 1984). This method is rapid, gives stable probes and can be used to label double-stranded and single-stranded DNA and RNA.

Antisera for specific DNA-RNA hybrids have been used for direct localisation of specific DNA sequences of genes coding for 18S and 28S ribosomal RNA (Rudkin & Stollar, 1977; Van Prooijen-Knegt et al., 1982). Application of this method is limited because the RNA-DNA hybrid must have a characteristic secondary structure which can elicit a specific antigenic response.

Finally, non-enzymatic incorporation of dinitrophenol (Dnp) into nucleic acids has been achieved by incubation with 2,4-dinitrobenzaldhyde at pH12. The Dnp hapten can then be visualised by indirect immunohistochemistry (Schroyer & Nakane, 1983).

Reporter molecules

A wide variety of fluorescent labels have been employed which include all those used for immunocytochemistry, for example, fluorescein, rhodamine, DAPI and Texas Red. Peroxidase and the more sensitive alkaline phosphatase conjugates are reliable alternative labels and can be visualised by standard cytochemical methods. Fluorescent labels give low background and very good resolution but have lower sensitivity than enzyme labels. Cell structures are also more difficult to interpret. Enzyme conjugates are sensitive and easily interpreted, without recourse to an expensive fluorescence microscope. However, because of the increase in the number of steps in the visualisation stage and the possibility of endogenous enzyme activity these methods can give higher background.

Electron microscopes clearly cannot visualise fluorescent or most enzyme labels (with the exception of peroxidase) but the advent of reliable colloidal gold conjugates allows very sensitive visualisation at both light and ultrastructural levels in plants (McFadden, 1989) and animals (Hutchinson et al., 1982; Binder, 1986, 1987). Sensitivity of gold probes, at light level, may be greatly increased; either, chemically by means of silver enhancement of the colloidal gold image (Holgate et al., 1983) or physically by Reflection

Contrast Microscopy (van der Ploeg & Van Duijn, 1979; Landegent *et al.*, 1985*a*, Cremers *et al.*, 1987).

In situ hybridisation protocol

Tissue pretreatments

Irrespective of the labelling method employed, tissues must be treated in an appropriate way before hybridisation can successfully be performed. The mains goals for successful *in situ* hybridisation are:

1. to prevent loss of target nucleic acids
2. to preserve tissue morphology
3. to allow penetration of probe.

Often, a desire to fulfil goals one and two are at odds with probe penetration. There are no definitive rules for tissue pretreatments and optimal conditions must be derived empirically for each tissue/probe combination.

Tissue preparations appropriate for *in situ* hybridisation
1. Fresh frozen cryostat sections
2. Perfusion or immersion fixed, frozen material
3. Fixed paraffin-embedded tissue
4. Cultured cells or smears
5. Fixed, plastic-embedded material for E.M.

Opinion differs as to whether tissue should be frozen before fixation or vice versa; and again depends upon the target and tissue. When localising mRNAs it may be good practice to fix first in order to inhibit the ubiquitous degradative RNAse enzymes and also reduce target diffusion. Best results are obtained for mRNA localisation if fixed tissue is immersed in sucrose solution prior to freezing.

Fixation

Fixation reduces nucleic acid loss, preserves morphology but reduces tissue permeability and therefore choice of fixative type, strength and duration is all important.

Fixatives routinely used for *in situ* hybridisation

Precipitators – e.g. acetone, ethanol/acetic acid (Carnoy's)
or
Crosslinkers – e.g. Formalin, para-formaldehyde, glutaraldehyde

Non-radioisotopic in situ hybridisation

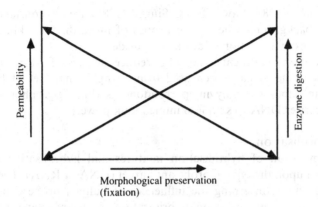

Figure 7. Fixation and enzyme digestion pretreatments are opposing factors in terms of attaining good morphological preservation and permeability of tissue sections.

Precipitating fixatives give best penetration but permit loss of small target nucleic acids and give relatively poor morphology. By comparison, the crosslinkers preserve morphology and target but reduce penetration by probes (McAllister & Rock, 1985). More often than not fixation will reduce probe penetration and a secondary permeabilisation step is necessary.

Prevention of section loss

Section loss is a consequence of the prolonged nature of *in situ* hybridisation procedure. This may be avoided by subbing slides prior to cutting sections. Many materials have been used such as gelatin, chrome–alum gelatin, Elmers glue and more recently poly-L-lysine and aminopropylethoxysilane.

Permeabilisation

This is usually an enzymic digestion with, for example, Proteinase K, Pronase or a pepsin/HCl. Again, fixation and permeabilisation work in opposition as can be seen in Figure 7. Achievement of optimal conditions can only be measured in terms of the strength of the final hybridisation signal.

Prehybridisation treatments to decrease non-specific binding of probe

Electrostatic attractions between probes and charged proteins in tissues result in characteristic 'stickiness' which has been suggested as a source of nonspecific binding. Some workers wash slides and coverslips with 0.25% acetic anhydride to reduce probe binding to negatively charged glass

(Hayashi et al., 1978). Lawrence & Singer (1985) found considerable reduction in background when using probes of more than 1.5 kb, after washing sections in a solution of acetic anhydride.

Prehybridisation treatments may also reduce non-specific binding of probes. Various media have been used from complex mixtures of ficoll, bovine serum albumin, polyvinylpyrrolidone, sodium pyrophosphate, EDTA and carrier DNA to simple skimmed milk power.

Hybridisation

Formulation of hybridisation mixtures and hybridisation conditions depend upon the type of probe and target (DNA or RNA). There is extensive literature concerning the influences of buffers, probe concentration, time and temperature on hybridisation efficiency (Lawrence & Singer, 1985). As a rule of thumb 10–50 ng of biotinylated probe are used per section for in situ hybridisations and is applied in a volume of about 10–20 μl. If either probe or target are DNA then the duplexes must be denatured by heating for about 10 minutes to at least 95°C. Coates et al. (1987) showed that even higher temperatures (105°C in a domestic microwave oven), gave much better hybridisation. Standard buffers contain 50% formamide (can be varied to give required stringency), $2 \times$ SSC (standard saline citrate $1 \times$ SSC = 0.15 M sodium chloride, 0.015 M sodium citrate) and dextran sulphate. It is suggested that dextran sulphate increases the rate and signal amplification of hybridisation by volume exclusion effectively concentrating the probe (Wahl, Stern & Stark, 1979). For convenience rather than necessity, hybridisations are often carried out overnight, though 3 hours may be sufficient and, depending upon the required stringency, hybridisations can be performed between 37°C and 69°C. Carrier non-homologous DNA (e.g. salmon sperm DNA) is routinely added to occupy the majority of non-specific binding sites, leaving the smaller amount of labelled probe free to bind to specific target sequence.

Post-hybridisation washes

Post-hybridisation washes are crucial because in situ hybridisations are usually at low stringency resulting in an appreciable degree of non-specific hybridisation. Heterologous duplexes have lower T_ms than homologous duplexes and are therefore inherently less stable. As low ionic strength reduces T_m, non-specific duplexes are removed by using a series of decreasing salt strength washes at a temperature near to the T_m of the homologous duplex. The extent of stringent washing depends on a number of factors such as the degree of fixation, probe concentration and length and the actual hybrid stability in terms of T_m. In practice, stringent wash protocols are determined empirically in order to obtain the lowest background without loss of signal strength. Formamide may also be added to wash solutions.

Single-stranded probe non-specifically bound to tissue components other than nucleic acids can sometimes be reduced by digestion with an appropriate nuclease. RNAse digestion was used by Lynn et al. (1983) to remove non-specifically bound riboprobe, and non-specifically bound cDNA can be removed by S1 nuclease (Godard, 1983). Post-hybridisation digestions are particularly helpful with the inherently more 'sticky' RNA probes.

Visualisation
Clearly, visualisation protocols depend on the type of label used and are essentially the same as for immunohistochemistry.

The importance and range of controls for specificity
Specific reaction is the paramount criteria for any histochemical method and the complexity of the *in situ* hybridisation technique introduces several stages where nonspecific reactions may result. False signals may arise due to three main factors. Either, from sequence-independent binding of probe to nucleic acid, or, due to non-specific binding to other tissue components, or finally as artefacts of label visualisation. Consequently, the range and number of controls is more extensive than any other histochemical method.

Controls of specificity of hybridisation
1. Northern or Southern blot
2. Combination of *in situ* hybridisation and immunohistochemistry
3. Heterologous probes ($-$ve)
4. Hybridisation with different fragments of the specific sequence ($+$ve)
5. Prehybridisation of probe with specific cDNA or cRNA ($-$ve)
6. Hybridisation with non-specific vector consequences ($-$ve)
7. Pretreatment of sections with RNAse or DNAse before or after hybridisation as appropriate ($-$ve)
8. Competition with unlabelled probe ($-$ve)
9. Thermal stability of probe–target duplex
10. Positive tissue control
11. Negative tissue control.

Immunohistochemical localisation of a protein and its mRNA within the same cell type, preferably the same cell on a single section is strong evidence of specific hybridisation of the probe (Brahic, Haase & Cash, 1984; Näher, Petzoldt & Sethi, 1988; Shivers *et al.*, 1986; Jirikowski *et al.*, 1988). Heterologous probes should, wherever practicable, be of similar length and

Multiple oligonucleotide control

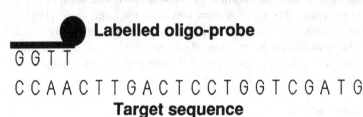

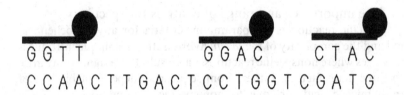

= 3 X INCREASE IN SIGNAL INTENSITY

Figure 8. Multiple oligonucleotide probes, synthesised as complementary to different sectors of the same gene sequence, can be used to confirm specific hybridisation if amplification of signal intensity results.

GC content to the test probe, a condition more easily fulfilled for riboprobes and synthetic oligonucleotides where sense strand probes may be used. Control 4 is particularly appropriate to the use of oligonucleotide probes. Specific hybridisation is indicated if three different oligonucleotides, complementary to different sectors of the same target gene or RNA, give the same cellular localisation. Multiple oligonucleotide probes hybridised on the same section should result in proportionate increase in signal strength (Fig. 8). Pre-hybridisation of the probe with complementary unlabelled cDNA should give negative results. Hybridisation with non-specific vector sequences is a convenient way of indicating the availability of non-specific binding sites in the section. Digestion of nucleic acids using broad spectrum DNAses or RNAses can establish whether a signal results from binding to nucleic acid sequences rather than to other molecules such as proteins. For example, if a DNA sequence is the target, pre-hybridisation digestion with DNAse should remove all signal, whereas RNAse will have no effect. When

Probe competition control

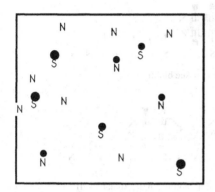 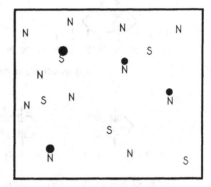

● LABEL   S = SPECIFIC   N = NON-SPECIFIC

Figure 9. Competition between labelled and unlabelled probes can be used as a negative control for hybridisation specificity. Specific hybridisation is indicated if the signal is attenuated.

using cDNA probes to localise mRNA, a negative result after RNAse treatment would be a valid indication of specific hybridisation. However, a problem of interpretation may arise when using riboprobes because RNAse is difficult to remove or inactivate and negative results may be due simply to digestion of the riboprobe by residual RNAse. One might instead use Micrococcal nuclease which completely digests intracellular RNA but is inactivated by EDTA during subsequent steps (Williamson, 1988). Competition between labelled and unlabelled probe can indicate the degree of non-specific binding (Fig. 9) as it is likely that there are more non-specific binding sites than specific. Therefore, in the presence of an excess of unlabelled probe, the rarer specific binding sites are more likely to be occupied by an unlabelled probe thus attenuating the specific signal. Probe–target stability may also be studied from melting curves, measured in terms of changing signal strength, and a value for T_m obtained. This may be compared either with a calculated T_m or with T_m derived from model hybridisation systems such as slot blots. Heterologous hybrids would give loss of signal at temperatures lower than T_m (Cox et al., 1984). It is always desirable to include positive and negative tissue controls but this may not be possible if you are using a new probe.

The controls described above are for determination of probe specificity but the type and complexity of the reporter steps determine which secondary

Blocking of endogenous avidin binding sites

Non-specific signal due to tissue biotin

Block non-specific binding with an avidin–biotin complex (unlabelled)

Avidin   Unlabelled biotin   Labelled biotin

Figure 10. Many tissues contain appreciable amounts of endogenous biotin. Endogenous non-specific avidin binding sites can be blocked successfully by treating sections with unlabelled avidin and biotin prior to hybridisation.

controls must be used. Many of these are similar to those employed for immunohistochemistry.

Secondary controls include:
1. Block non-specific antibody binding sites using bovine serum albumin or milk powder
2. Block endogenous biotin (Fig. 10)
3. Inhibition of endogenous enzyme activity.

Endogenous biotin can cause considerable non-specific staining (Bannerjee & Pettit, 1984; Wood & Warnke, 1981). Attenuation of non-specific biotin–avidin binding has been achieved by using $4 \times$ SSC washes, surprisingly phosphate buffered saline has been shown to increase background staining with avidin conjugates (Singer, Lawrence & Rashtchian, 1987).

Example method
Tissue preparation
1. Freeze tissue in isopentane cooled in nitrogen.
2. Cut cryostat sections 6–8 μm and place on poly-L-lysine coated slides.

Fixation
1. Fix sections for 30 min in 4% paraformaldehyde in Buffer 1(0.1 M phosphate buffered pH 7.2–7.4, 0.1 M sucrose).
2. Wash in distilled H_2O.
   Treat slides with 0.25% acetic anhydride for 5 min.
3. Dehydrate in alcohol series and store at $-20°C$.
4. Formalin-fixed paraffin embedded sections were dewaxed at this stage and rehydrated before use.
   Wash sections in Buffer 1.
5. Permeablise tissue with 0.1 mg/ml pepsin/0.1 M HCl or 0.5 mg/ml proteinase K at 37°C for 15 min.
6. Wash extensively in 0.1 M Tris/HCl (pH 7.2) and then air dry. Our probe, pY353/B, is a 1.5 kb cDNA insert in pUC9 produced by PCR (25 cycles 95°C for 1.5 min, 55°C for 1.5 min and 72°C for 4 min. Probes were either biotinylated during PCR using incorporation of biotin-11-UTP or were subsequently nick-translated with biotin-11-dUTP to give 200–300 base pair fragments.

Pre-hybridisation
Immerse in $2 \times$ SSC briefly.
Incubate in $2 \times$ SSC + 45% formamide for 10 min.
Pre-hybridise with hybridisation buffer 2 (45% formamide, $2 \times$ phosphate buffered saline pH 7.2 (PBS), $5 \times$ Denhardt's, 1 mM EDTA, 100 mg/ml salmon sperm DNA) without probe for 1 hour.
Dip briefly in $2 \times$ SSC before adding hybridisation mixture.

Hybridisation

7. Pipette on hybridisation mixture containing 2 μg/ml pY353/B probe (with or without 10% dextran sulphate).
8. Cover with coverslip and seal; denature probe and target sequence at 90–95°C for 10 min.
9. Hybridise overnight at 37°C.

Post-hybridisation
10. Non-specifically bound probe may be removed by washing in PBS/0.05% Triton X-100 and decreasing salt strength washes before blocking with 5% bovine serum albumin (BSA) or low fat dried milk for 5 min.
11. Rinse sections in 0.01 M Tris/HCl (pH 7.5).

Visualisation
12. Incubate sections with avidin–alkaline phosphatase conjugate (2 mg/ml) room temp 20–30 min.
13. Wash several times with 0.01 M Tris/HCl (pH 7.5).
14. Wash in enzyme detection buffer (0.1 M Tris/HCl pH 9.5, 0.1 M NaCl and 50 mM $MgCl_2$ for 1 h.
15. Localise sites of hybridisation using chromogenic substrate medium containing Nitroblue tetrazolium (NBT) (75 mg/ml in 70% dimethylformamide) and bromo-chloro-indolyl phosphate (BCIP) (50 mg/ml in 70% dimethylformamide) in 7.5 ml of detection buffer.
Incubate in the dark for 30 min–3 h at 22–37°C.

Applications of non-radioisotopic *in situ* hybridisation

The extensive range of applications which presently use non-radioisotopic *in situ* hybridisation includes: bacterial detection, viral detection, chromosomal aberrations, chromosome mapping, mRNA localisation, prenatal sex determination and oncogene detection.

Hybridocytochemistry has begun to address previously insoluble problems in the field of gene expression. Immunocytochemistry may be able to specifically localise an antigen within tissue sections but this is not necessarily the site of protein synthesis as many peptides are synthesised and secreted in one cell type and endocytosed by another. Analysis of gene expression by immunohistochemistry alone is unreliable because high concentrations of antigen may not necessarily indicate very active synthesis. This is because the concentration of antigen is dependent not only on synthetic rate but also on rates of degradation and secretion. Detection of mRNA templates by *in situ* hybridisation gives a more direct indication of the rate of gene expression. Most early reports describe localisation of mRNA using radiolabelled probes (Lawrence & Singer, 1985; Breborowicz & Tanaoki, 1985; Gresik, Gubits & Barka, 1985; McCabe *et al.*, 1986). Relatively few mRNA studies have used non-radiolabelled probes, most of these have used biotinylated probes (Bresser & Evinger-Hodges, 1987; Larsson *et al.*, 1988; Lawrence & Singer, 1986; Smith *et al.*, 1986; Webster *et al.*, 1987; Zabel & Schäfer, 1988; Arai *et al.*, 1988). Dirks and co-authors (1989) compared several types of non-radiolabelled and radiolabelled cDNA and oligonucleotide probes in their investigation of the model neuroendocrine system of *Lymnea stagnalis*. They optimised the *in situ* protocol in terms of fixation, permeabilisation and hybridisation for the use of biotinylated, sulphonated and AAF-linked probes. They make no comment on relative sensitivity but mention the greater resolution of the non-radiolabelled probes. Biotinylated riboprobes have been used to compare the distribution of calcitonin and calcitonin-gene-related peptide mRNAs with protein products in rat parafollicular cells (Zabel & Schäfer, 1988). Combined *in situ* hybridisation and immunocytochemical studies provide the greatest insight into protein synthesis and have again been carried out primarily using radiolabelled probes (Shivers *et al.*, 1986; Steel *et al.*, 1988; Chan-Palay *et al.*, 1988).

Viral detection in diagnostic pathology (e.g. Chapters 12, 13) is another area where non-radioisotopically labelled probes are in the forefront. This is especially true for identification of latent viral states as quiescent virus do not synthesise proteins and cannot be localised by immunocytochemistry. Brigati *et al.* (1983) using biotinylated DNA probes labelled by nick translation detected parvovirus, polyomavirus, herpes simplex virus and adenovirus in cultured mouse cells and human autopsy material. Non-radioisotopic localisations of cytomegalovirus in human lung (Przepiorka & Myerson, 1986; Niedobitek *et al.*, 1988; Unger *et al.*, 1986) and human liver (Naoumov *et al.*, 1988), hepatitis B in human liver (Naoumov *et al.*, 1988), human papilloma virus in human archival cervix material (Burns *et al.*, 1988) and herpes in tonsils of patients with herpes encephalitis (Burns *et al.*, 1986) have been reported. The use of biotinylated probes in viral diagnostic studies may reflect the clinicians' desire for rapid results and also the applicability of non-radioisotopic techniques to routine laboratories. High titres of viral particles are often found within infected cells, hence localisation is not hindered by the lower sensitivity of non-radiolabelled probes. Chronic or latent viral infections may however give rise to very low copy number within infected cells. In response to this, Burns *et al.* (1987) developed protocols using biotinylated probes for the detection of low copy number human papilloma virus in routinely fixed wax-embedded sections. Using an avidin–alkaline phosphatase complex the authors detected < 10 copies of HPV per cell. Possibly of greater interest was the discovery of previously unidentified, low copy number viral particles in suprabasal cells of condylomas.

Biotinylated probes are the most commonly used for viral diagnosis but a range of other labelling methods have been employed to try and improve sensitivity and speed. AAF-modified cDNA probes were used to localise cytomegalovirus infected mouse liver cells (Raap *et al.*, 1988). Niedobitek *et al.*, (1988) successfully used nick-translated bromodeoxyuridine (Brd–Urd)-containing DNA probes for the localisation of cytomegalovirus in human lung. The authors claim similar levels of sensitivity to biotin-labelled probes. Human immunodeficiency virus (HTLV-III) was detected in lymph nodes of lymphadenopathy patients by sulphonated DNA probes (Pezzella *et al.*, 1987). The sensitivity of this method, in terms of viral copy number detection, remains to be determined.

The identification of chromosomal rearrangements and physical gene mapping, once thought to be the sole realm of radioisotopic detection, are now possible using non-radioisotopic labels (Landegent et al., 1985a,b) Hopman, Wiegant & Van Duijn (1987) were able to detect unique human DNA fragments as small as 15.6 kb in metaphase chromosomes using mercurated probes. Mercurated probes have also been used to study numerical chromosome aberrations in solid bladder tumors (Hopman et al., 1988). Direct visualisation of single copy genes is also possible using biotinylated probes detected by antibody–peroxidase or colloidal gold complexes (Garson, van den Berghe & Kemshead, 1987; Bhatt et al., 1988).

Future developments

This is a very exciting time to be involved in *in situ* hybridisation as we are on the eve of a revolution in its development. Fifteen years have seen a change from an offshoot of molecular biology into one of the most powerful research tools available to the cell biologist. The three main driving forces behind this change are the ever-increasing supply of probes from molecular genetic research, the development of more rapid and reliable probe production methods and the continuing development of increasingly sensitive and simple non-radioactive labelling methods. Continued development of non-radioactive labelled methods is the only way in which *in situ* hybridisation can follow in the footsteps of immunocytochemistry to become a standard method in routine research and diagnostic laboratories.

It is difficult to look into the crystal ball and single out particular areas of research which will derive greatest benefit from new developments but, clearly, the fields of gene expression, chromosomal analysis and viral diagnosis will remain in the forefront. I think that biotin–avidin systems will remain the most popular method for the foreseeable future due to the ease of labelling and the relatively sensitive detection systems available. The great potential of riboprobes will also be realised when mercurated or sulphonated ribonucleotides become available for use in *in vitro* transcription. There is also the intriguing possibility of oligo-riboprobes for *in situ* hybridisation (Denny et al., 1988). The use of non-radioisotopic probes for use in flow cytometry, already successful in chromosome analysis (Trask et al., 1988) will rapidly expand.

Increasing numbers of new non-radioisotopic labelling methods, developed initially for blotting methods are described in the literature. A very recent study of hepatitis B DNA detection describes a new method for attachment of hapten moieties on to DNA involving reaction of DNA with N-bromosuccinimide at alkaline pH, resulting in bromination of guanine and cytosine residues (Keller et al., 1988). DNA has also been modified by aliphatic amino groups (Haralambis, Choi & Tregear, 1987; Adarichev et al.,

1987), chelate group substituted psoralen/europium (Oser, Roth & Valet, 1988) and bromovinyl groups (Sági et al., 1987) for use as non-radioactive probes.

An early indication of the coming revolution may be seen in the increasing availability of commercially available cDNA probes to targets such as cytomegalovirus.

Finally, we should now recognise the quantum leap which non-radioisotopic labelling methods have taken in recent years by choosing a new and more appropriate name for this family of techniques. It is patently illogical to call these methods by a negative term which in no way describes their chemical or physical basis. All these methods depend upon direct or indirect binding of a reporter molecule to nucleic acid probes. I propose that the name should reflect this and they should be called Affinity-Complex Labelled Probes (ACLP).

References

Adarichev, V. A., Dymshits, G. M., Kalaachikov, S. M., Pozniakov, P. I. & Salganik, R. I. (1987). DNA carrying aliphatic amino groups and the use of its fluorescent derivative as a probe in molecular hybridisation. *Bioorganischeskaia Khimiia (Moskva)*, **13 (8)**, 1066–9.

Angerer, L. M., Stoler, M. H. & Angerer, R. C. (1987). *In situ* hybridisation with RNA probes – an annotated recipe. In In situ *hybridization – applications to neurobiology* (eds Valentino, K. L., Eberwine, J. H. & Barchas, J. D.) pp. 71–96, Oxford University Press.

Arai, H. Emson, P. C., Agrawal, S. Christodoulou, C. & Gait, M. J. (1988). *In situ* hybridization histochemistry: localization of vasopressin mRNA in rat brain using biotinylated oligonucleotide probe. *Brain Research*, **464 (1)**, 63–9.

Bannerjee, D. & Pettit, S. (1984). Endogenous avidin–binding activity in human lymphoid tissue. *Journal of Clinical Pathology*, **37**, 223–5.

Bauman, J. G., Wiegant, J. & van Duijn, P. (1981). Cytochemical hybridization with fluorochrome-labeled RNA. *Journal of Histochemistry and Cytochemistry*, **29**, 238–40.

Bhatt, B., Burns, J., Flannery, D. & O'D McGee, J. (1988). Direct visualization of single copy genes on banded metaphase chromosomes by nonisotopic *in situ* hybridization. *Nucleic Acids Research*, **16 (9)**, 3951–61.

Binder, M. (1987). *In situ* hybridisation at the electron microscope level. *Scanning Microscopy*, **1**, 331–8.

Binder, M., Tourmente, S., Roth, J., Renaud, M. & Gehring, W. J. (1986). *In situ* hybridization at the electron microscope level: localization of transcripts on ultrathin sections of Lowicryl K4M-embedded tissue using biotinylated probes and protein A-gold complexes. *Journal of Cell Biology*, **102**, 1646–53.

Bollum, F. J. (1974). Terminal deoxynucleotide transferase. In *The*

Enzymes. (ed. Boyer, P. D.) vol. 10, 3rd edn, pp. 145–69, New York: Academic Press.

Boultwood, J., Wynford-Thomas, D., Wynford-Thomas, V. & Williams, E. D. (1986). Development of high sensitivity nucleic acid probes for in situ hybridization. *Journal of Pathology* Abstract, **148**, 61.

Brahic, M., Haase, A. T. & Cash, E. (1984). Simultaneous *in situ* detection of viral RNA and antigens. *Proceedings of the National Academy of Sciences, USA*, **81**, 5445–8.

Breborowicz, J. & Tamaoki, T. (1985). Detection of messenger RNAs of alphafetoprotein and albumin in a human hepatoma cell line by *in situ* hybridization. *Cancer Research*, **45**, 1730–6.

Bresser, J. & Evinger-Hodges, M. J. (1987). Comparison and optimization of *in situ* hybridization procedures yielding rapid, sensitive mRNA detections. *Genetic Analysis & Technology*, **4**, 89–92.

Brigati, D. J., Myerson, D., Leary, J. L., Spalholz, B., Travis, S. Z., Fong, C. K. Y., Hsiung, G. D. & Ward, D. (1983). Detection of viral genomes in cultured cells and paraffin-embedded tissue sections using biotin-labeled hybridisation probes. *Virology*, **126**, 32–50.

Buongiorno-Nardelli, M. & Amaldi, F. (1969). Autoradiographic detection of molecular hybrids between rRNA and DNA in tissue sections. *Nature, (London)*, **225**, 946–7.

Burns, J., Chan, V. T. W., Jonasson, J. A., Fleming, K. A., Taylor, S. & O'D McGee, J. (1985). Sensitive system for visualising biotinylated DNA probes hybridised *in situ*: rapid sex determination of intact cells. *Journal of Clinical Pathology*, **38**, 1085–92.

Burns, J., Graham, A. K., Frank, C., Fleming, K. A., Evans, M. F. & O'D McGee, J. (1987). Detection of low copy human papilloma virus DNA and mRNA in routine paraffin sections of cervix by non-radioisotopic *in situ* hybridisation. *Journal of Clinical Pathology*, **40**, 858–64.

Burns, J., Graham, A. K. & O'D McGee, J. (1988). Non-isotopic detection of *in situ* nucleic acid in cervix: an updated protocol. *Journal of Clinical Pathology*, **41**, 897–9.

Burns, J., Redfern, D. R. M., Esiri, M. M. & O'D McGee, J. (1986). Human and viral gene detection in routine paraffin embedded tissue by *in situ* hybridisation with biotinylated probes: viral localisation in herpes encephalitis. *Journal of Clinical Pathology*, **39**, 1066–73.

Casey, J. & Davidson, N. (1977) Rates of formation and thermal stabilities of RNA:RNA and DNA:DNA duplexes at high concentrations of formamide. *Nucleic Acids Research*, **4**, 1539–53.

Chan-Palay, Y., Yasargil, G., Hamid, Q., Polak, J. M., & Palay, S. L. (1988). Simultaneous demonstrations of neuropeptide Y gene expression and peptide storage in single neurons in the human brain. *Proceedings of the National Academy of Sciences, USA*, **85 (9)**, 3213–15.

Coates, P. J., Hall, P. A., Butler, M. G. & D'Ardenne, A. J. D. (1987). Rapid technique of DNA–DNA *in situ* hybridisation on formalin fixed tissue sections using microwave irradiation. *Journal of Clinical Pathology*, **40**, 865–9.

Cox, K. H., DeLeon, D. V., Angerer, L. M. & Angerer, R. C. (1984). Detection of mRNAs in sea urchin embryos by *in situ* hybridisation using asymmetric RNA probes. *Developmental Biology*, 101, 485–502.

Cremers, A. F., Jannsen in de Wal, N., Wiegant, J., Dirks, R. W., Weisbeck, P., van der Ploeg, M. & Landegant, J. E. (1987). Non-radioactive *in situ* hybridization: a comparison of several immunocytochemical detection systems using reflection contrast and electron microscopy. *Histochemistry*, 86 (6), 609–15.

Crum, C. P., Nuovo, G., Friedman, D & Silverstein, S. J. (1988). A comparison of biotin and isotope-labeled ribonucleic acid probes for *in situ* detection of HPV-16 ribonucleic acid in genital precancers. *Laboratory Investigation*, 58 (3), 354–9.

Denny, P., Hamid, Q., Krause, J. E., Polak, J. M. & Legon, S. (1988). Oligoriboprobes: tools for *in situ* hybridization. *Histochemistry*, 89, 481–3.

Dirks, R. W., Raap, A. K., van Minnen, J., Vreugdenhil, E., Smit, A. B. & van der Ploeg, M. (1989). Detection of mRNA molecules coding for neuropeptide hormones of the pond snail *Lymnea stagnalis* by radioactive and non-radioactive *in situ* hybridization: a model study for mRNA detection. *Journal of Histochemistry and Cytochemistry*, 37 (1), 7–14.

Feinberg, A. P. & Vogelstein, B. (1983). A technique for radiolabelling DNA restriction endonuclease fragments to a high specific activity. *Analytical Biochemistry*, 132, 6–13.

Feinberg, A. P. & Vogelstein, B. (1984). A technique for radiolabelling DNA restriction endonuclease fragments to a high specific activity. *Addendum Analytical Biochemistry*, 137, 266–7.

Forster, A. C., McInnes, J. L., Skingle, D. C. & Symons, R. H. (1985). Non-radioactive hybridization probes prepared by the chemical labelling of DNA and RNA with a novel reagent photobiotin. *Nucleic Acids Research*, 13, 743–61.

Gall, G. & Pardue, M. L. (1969). Formation and detection of RNA–DNA hybrid molecules in cytological preparations. *Proceedings of the National Academy of Sciences, USA*, 63, 378–81.

Gall, G. & Pardue, M. L. (1971). Nucleic acid hybridization in cytological preparations. *Methods in Enzymology*, 38, 470–80.

Garson, J. A., van den Berghe, J. A. & Kemshead, J. T. (1987). Novel non-isotopic *in situ* hybridization technique detects small (1 kb) unique sequences in routinely G-banded human chromosomes: fine mapping of N-*myc* and beta-*NGF* genes.

Godard, C. M. (1983). Improved method for detection of cellular transcripts by *in situ* hybridization. *Histochemistry*, 77, 123–31.

Gresik, E. W., Gubits, R. M. & Barka, T. (1985). *In situ* localization of mRNA for epidermal growth factor in the submandibular gland of the mouse. *Journal of Histochemistry and Cytochemistry*, 33, 1235–40.

Gu, J., Linnoila, R. I., Siebel, N. L., Gazdar, A. F., Minna, J. G. Brooks, B. J., Hollis, G. F. & Kirsch, I. R. (1988). A study of *myc*-related gene expression in small cell lung cancer by *in situ* hybridization. *American Journal of Pathology*, 132, 13–17.

Haralambis, J., Choi, M. & Tregear, G. W. (1987). Preparation of base-modified nucleosides suitable for non-radioactive label attachment and their incorporation into synthetic oligodeoxyribonucleotides. *Nucleic Acids Research*, **15 (12)**, 4857–76.

Harris, N. & Croy, R. R. D. (1986). Localization of mRNA for pea legumin: *in situ* hybridization using a biotinylated DNA probe. *Protoplasma*, **130**, 57–67.

Hayashi, S., Gillam, I. C., Delaney, A. D. & Tener, G. M. (1978). Acetylation of chromosome squashes of *Drosophila melanogaster* decreases the background in autoradiographs from hybridisation with [^{125}I]-labeled RNA. *Journal of Histochemistry and Cytochemistry*, **26 (8)**, 677–9.

Höfler, H. (1987). What's new in 'In Situ Hybridisation'. *Pathology Research and Practice*, **182**, 421–430.

Holgate, C., Jackson, P., Cowen, P. & Bird, C. (1983). Immunogold–silver staining: new method of immunostaining with enhanced sensitivity. *The Journal of Histochemistry and Cytochemistry*, **31**, 938–46.

Hopman, A. H. N., Ramaekers, F. C. S., Raap, A. K., Beck, J. L. M., Devilee, P., van der Ploeg, M. & Vooijs, G. P. (1988). *In situ* hybridisation as a tool to study numerical chromosomes aberrations in solid bladder tumors. *Histochemistry*, **89**, 307–16.

Hopman, A. H. N., Wiegant, J., Tesser, G. I., van Duijn, P. (1986). A nonradioactive *in situ* hybridization method based on mercurated nucleic acid probes and sulfhydryl–hapten ligands. *Nucleic Acids Research*, **14**, 6471–88.

Hopman, A. H. N., Wiegant, J. & van Duijn, P. (1987). Mercurated Nucleic acid probes. A new principle for non-radioactive *in situ* hybridisation. *Experimental Cell Research*, **169**, 357–68.

Hutchinson, N. J., Langer–Safer, P. R., Ward, D. & Hamkalo, B. A. (1982). *In situ* hybridization at the electron microscope level: hybridi detection by autoradiography and colloidal gold. *Journal of Cell Biology*, **95**, 609–18.

Jirikowski, G. F., Ramalho-Ortigao, F. & Seliger, H. (1988). *In situ* hybridization with complementary synthetic oligonucleotide and immunocytochemistry: a combination of methods to study transcription and secretion of oxytocin by hyperthalamic neurons. *Molecular and Cellular Probes*, **2**, 59–64.

Keller, G. H., Cumming, C. U., Huang, D. P., Manak, M. M. & Ting, R. (1988). A chemical method for introducing haptens onto DNA probes. *Analytical Biochemistry*, **170 (2)**, 441–50.

Landegent, J. E., Jansen in der Wal, N., Baan, R. A., Hoeijmakers, J. H. J. & van der Ploeg, M. (1984). 2-Acetylaminofluorene-modified probes for the indirect hybridocytochemical reaction of specific nucleic acid sequences. *Experimental Cell Research*, **153**, 61–72.

Landegent, J. E., Jansen in de Wal, N., van Ommen, G-J. B., Baas, F., de Vilder, J. J. M., van Duijn, P. & van der Ploeg, M. (1985*a*). Chromosomal localization of a unique gene by non-autoradiographic *in situ* hybridization. *Nature (London)*, **317**, 175–7.

Landegent, J. E., Jansen in de Wal, N., Ploem, J. S. & van der Ploeg, M. (1985b). Sensitive detection of hybridocytochemical results by means of reflection-contrast microscopy. *Journal of Histochemistry and Cytochemistry*, **33 (2)**, 1241–6.

Langer, P. R., & Ward, D. (1981). A sensitive immunological method for in situ gene mapping. In *Developmental Biology Using Purified Genes* (ed. Brown, D. D.) vol 23, pp. 647–58, New York: Academic Press.

Langer, P. R., Waldrop, A. A. & Ward, D. C. (1981). Enzymatic synthesis of biotin-labeled polynucleotides: novel nucleic acid affinity probes. *Proceedings of the National Academy of Sciences, USA*, **78 (11)**, 6633–7.

Langer-Safer, P. R., Levine, M. & Ward, D. (1982). Immunological method for mapping genes on *Drosophila* polytene chromosomes. *Proceedings of the National Academy of Sciences, USA*, **79**, 4381–5.

Larsson, L-I., Christensen, T. & Dalboge, H. (1988). Detection of proopiomelanocortinin mRNA by *in situ* hybridization, using a biotinylated oligodeoxynucleotide probe and avidin-alkaline phosphatase histochemistry. *Histochemistry*, **89**, 109.

Lawrence, J. B. & Singer, R. H. (1985). Quantitative analysis of *in situ* hybridisation for the detection of actin gene expression. *Nucleic Acids Research*, **13 (5)**, 1777–99.

Lawrence, J. B. & Singer, R. H. (1986). Intracellular localization of messenger RNAs for cytoskeletal proteins. *Cell*, **45**, 407–15.

Lewis, M. E., Sherman, T. G. & Watson, S. J. (1985). *In situ* hybridisation with synthetic oligonucleotides: strategies and methods. *Peptides*, **6, Suppl. 2**, 75–87.

Liesi, P., Julien, J-P., Vilja, P., Grosveld, F. & Rechardt, L. (1986). Specific detection of neuronal cell bodies: *in situ* hybridization with a biotin-labeled neurofilament cDNA probe. *Journal of Histochemistry and Cytochemistry*, **34 (7)**, 923–6.

Lo, Y-M. D., Mehal, W. Z. & Fleming, K. A. (1988). Rapid production of vector free biotinylated probes using the polymerase chain reaction. *Nucleic Acids Research*, **16 (17)**, 8719.

Lynn, D. A., Angerer, L. M., Bruskin, A. M., Klein, W. H. & Angerer, R. C. (1983). Localization of a family of mRNAs in a single cell type and its precursors in sea urchin embryos. *Proceedings of the National Academy of Sciences, USA*, **80**, 2656–60.

Maniatis, T., Fritsch, E. F. & Sambrook, J. (1982). Molecular cloning: a laboratory manual. *Cold Spring Harbor Laboratory, Cold Spring Harbor, NY*.

Mannuelidis, L., Langer-Safer, P. R. & Ward, D. (1982). High resolution mapping of satelite DNA using biotin-labeled DNA probes. *Journal of Cell Biology*, **95**, 619–25.

McAllister, H. A. & Rock, D. L. (1985). Comparative usefulness of tissue fixatives for *in situ* viral nucleic acid hybridization. *Journal of Histochemistry and Cytochemistry*, **33 (10)**, 1026–32.

McCabe, J. T., Morrel, J. I., Ivell, R., Schmale, H., Richter, D. & Pfaff, D. W. (1986). *In situ* hybridization technique to localize rRNA and mRNA in mammalian neurons. *Journal of Histochemistry and Cytochemistry*, **34** (1), 45–50.

McFadden, G. I. (1989). In situ hybridisation in plants: from macroscopic to ultrastructural resolution. *Cell Biology International Reports*, **13**, 3–20.

Morley, D. J. & Hodes, M. E. (1988). Amylase expression in human parotid neoplasms: evidence by *in situ* hybridization for the lack of transcription of the amylase gene. *Journal of Histochemistry and Cytochemistry*, **36 (5)**, 487–491.

Näher, N., Petzoldt, D. & Sethi, K. K. (1988). Evaluation of non-radioactive in situ hybridisation method to detect *Clamydia trachomatis* in cell culture. *Genitourinary Medicine*, **64**, 162–4.

Naoumov, N. V., Alexander, G. J. M., Eddleston, A. L. W. F. & Williams, R. (1988). *In situ* hybridization in formalin fixed, paraffin wax embedded liver specimens: method for detecting human and viral DNA using biotinylated probes. *Journal of Clinical Pathology*, **41**, 793–8.

Niedobitek, G., Finn, T., Herbst, H., Bornhöft, G., Gerdes, J. & Stein, H. (1988). Detection of viral DNA by *in situ* hybridization using bromodeoxyuridine-labeled DNA probes. *American Journal of Pathology*, **131**, 1–4.

Oser, A., Roth, W. K. & Valet, G. (1988). Sensitive non-radioactive dot–blot hybridisation using DNA probes labelled with chelate group-substituted psoralen and quantitative detection by europium ion fluorescence. *Nucleic Acids Research*, **16 (3)**, 1181–96.

Pezzella, M., Pezzella, F., Galli, C., Macchi, B., Verani, P., Sorice, F. & Baroni, C. D. (1987). *In situ* hybridization of human immunodeficiency virus (HTLV-III) in cryostat sections of lymph nodes of lymphadenopathy syndrome patients. *Journal of Medical Virology*, **22**, 135–42.

Przepiorka, D. & Myerson, D. (1986). A single-step silver enhancement method permitting rapid diagnosis of cytomegalovirus infection in formalin-fixed, paraffin-embedded tissue sections by *in situ* hybridisation and immunoperoxidase detection. *Journal of Histochemistry and Cytochemistry*, **34 (12)**, 1731–4.

Raap, A. K., Geelen, J. C., vander Meer, J. W., van de Rijke, F. M., van den Boogaart, P. & van der Ploeg, M. (1988). Sensitive non-radioactive dot–blot hybridization using DNA probes labelled with chelate group substituted psoralen and quantitative detection by europium ion fluorescence. *Nucleic Acids Research*, **16 (3)**, 1181–96.

Richardson, C. C. (1981). T4 polynucleotide kinase. In *The Enzymes* (ed. Boyer, P. D.) vol. 10, New York: Academic Press.

Rigby, P. J. W., Dieckmann, M., Rhodes, C. & Berg, P. (1977). Labelling deoxyribonucleic acid to a high specific activity *in vitro* by nick translation with DNA polymerase. *Journal of Molecular Biology*, **113**, 237–51.

Rudkin, G. T. & Stollar, B. D. (1977). High resolution detection of DNA–RNA hybrids *in situ* by indirect immunofluorescence. *Nature (London)*, **265**, 472–3.

Sági, J., De Clerq, E., Szemzö, A., Csárnyi, A., Kovács, T. & Ötvös, L. (1987). Incorporation of the carboxylic analogue of (*E*)-5-(2-bromovinyl)2'-deoxyuridine 5'-triphosphate into a synthetic DNA. *Biochemical and Biophysical Research Communications*, **147** (3), 1105–12.

Saiki, R. K., Bugawan, T. L., Horn, G. T., Mullis, K. B., & Erlich H. A. (1986). *Nature (London)*, **324**, 163–6.

Schroyer, K. R. & Nakane, P. K. (1983). Use of DNP-labeled cDNA for *in situ* hybridization. *Cytogenetics*, **97**, 377a.

Shivers, B. D., Harlan, R. E., Pfaff, D. W. & Schachter, B. S. (1986). Combination of immunocytochemistry and *in situ* hybridisation in the same tissue section of rat pituitary. *Journal of Histochemistry and Cytochemistry*, **34** (1), 39–43.

Singer, R. H., Lawrence, J. B. & Raschtian, R. N. (1987). Toward a rapid and sensitive *in situ* hybridization methodology using isotopic and non-isotopic probes. In In situ *hybridization: applications to neurobiology* (eds Valentino, K. L., Eberwine J. H., & Barchas J. D.), pp. 71–96, New York & Oxford: Oxford University Press.

Smith, G. H., Doherty, P. J., Stead, R. B., Gorman, C. M., Graham, D. E. & Howard, B. H. (1986). Detection of transcription and translation in situ with biotinylated molecular probes in cells transfected with recombinant DNA plasmids. *Analytical Biochemistry*, **156**, 17–24.

Steel, J. H., Hamid, Q., van Noorden, S., Jones, P., Denny, P., Burrin, J., Legon, S., Bloom, S. R. & Polak, J. M. (1988). Combined use of *in situ* hybridisation and immunohistochemistry for the investigation of prolactin gene expression in immature, pubertal, pregnant, lactating and ovariectomised rats. *Histochemistry*, **89** (1), 75–80.

Thomas, C. A. & Dancis, B. M. (1973). Ring stability. *The Journal of Molecular Biology*, **77**, 44–55.

Traincard, F., Ternynck, T., Dancin, A. & Avrameas, S. (1983). Une technique immunoenzymatique pour le mise en evidence de l'hybridation moleculaire entire acides nucleiques. *Annals of Immunology (Institute Pasteur)* **134D**, 399–405.

Trask, B., van den Engh, G., Pinkel, D., Mullikin, J., Waldman, F., van Dekken, H. & Gray, J. (1988). Fluorescence in situ hybridisation to interphase call nuclei in suspension allows flow cytometric analysis of chromosome content and microscope analysis of nuclear organisation. *Human Genetics*, **78** (3), 251–9.

Unger, E. R., Budgeon, L. R., Myerson, D. & Brigati, D. J. (1986). Viral diagnosis by *in situ* hybridization. *American Journal of Surgical Pathology*, **10** (1), 1–8.

van Prooijen-Knegt, A. C., van Hoek, J. F. M., Bauman, J. G. J., van Duijn, P., Wool, I. G. & van der Ploeg, M. (1982). *In situ* hybridisation of DNA sequences in human metaphase chromosomes visualised by an indirect fluorescent immunocytochemical procedure. *Experimental Cell Research*, **141**, 397–407.

van der Ploeg, M. & van Duijn, P. (1979). Reflection versus fluorescence.

Histochemistry, **62**, 227–32.
van der Ploeg, M., Landegent, J. E., Hopman, A. H. N. & Raap, A. K. (1986). Non-autoradiographic hybridocytochemistry. *Journal of Histochemistry and Cytochemistry*, **34**, 126–33.
Varndell, L. M., Polak, J. M., Sikri, K. L., Minth, C. D., Bloom, S. R. & Dixon, J. E. (1984). Visualisation of messenger RNA directing peptide synthesis by *in situ* hybridisation using a novel single-stranded cDNA probe. *Histochemistry*, **81**, 597–601.
Verdlov, E. D., Monastryskaya, G. S., Guskova, L. I., Levitan, E. J., Sheishenko, V. J. & Budowsky, E. J. (1974). Modification of cytidine residues with a bisulfite-*o*-methylhydroxylamine mixture. *Biochimica et Biophysica Acta*, **340**, 153–65.
Wahl, G. M., Stern, M. & Stark, G. R. (1979). Efficient transfer of large Dna fragments from agarose gels to diazobenzyloxymethyl-paper and rapid hybridization by using dextran sulfate. *Proceedings of the National Academy of Sciences, USA*, **76**, 3683–7.
Webster, H. F., Lamperth, L. K., Fovilla, J. T., Lemke, G., Tesin, D. & Manuelidis, L. (1987). Use of biotinylated probe and in situ hybridization for light and electron microscopic localization of Po mRNA in myelin-forming Schwann cells. *Histochemistry*, **86**, 441.
Williamson, D. J. (1988). Specificity of riboprobes for intracellular RNA in situ hybridisation histochemistry. *Journal of Histochemistry and Cytochemistry*, **36 (7)**, 811–13.
Wood, G. S. & Warnke, R. (1981). Suppression of endogenous avidin-binding activity in tissues and its relevance to biotin–avidin detection systems. *Journal of Histochemistry and Cytochemistry*, **29**, 1196–204.
Zabel, M. & Schäfer, H. (1988). Localization of calcitonin and calcitonin gene-related peptide mRNAs in rat parafollicular cells by hybridocytochemistry. *Journal of Histochemistry and Cytochemistry*, **36 (5)**, 543–6.

A. K. Raap, P. M. Nederlof, R. W. Dirks, J. C. A. G. Wiegant and M. van der Ploeg

Use of haptenised nucleic acid probes in fluorescent *in situ* hybridisation

General

With the advent in this decade of various nucleic acid labelling techniques which allow a non-radioactive visualisation of hybridisation results, the technique of *in situ* hybridisation, originally developed by Pardue & Gall (1969) and Jones & coworkers (John *et al.*, 1969) have become increasingly important in many biomedical fields. Next to the environmentally important feature of absence of radiation hazards, favourable properties of the non-radioactive techniques such as the high topological resolution attainable, the stability of the labelled nucleic acid probes and the capability to perform multiple hybridisations simultaneously, have contributed significantly to this development.

Haptenisation of probes

Essentially, two types of non-radioactive *in situ* hybridisation methods can be distinguished: direct methods in which the microscopic reporter molecules, e.g. fluorochromes (Bauman *et al.*, 1980) or enzymes (Renz & Kurz, 1984) are directly coupled to the nucleic acid probe molecules and the indirect techniques in which the nucleic acid probe is labelled with a hapten which, after *in situ* hybridisation, is detected by immunocytochemical means (Raap *et al.*, 1989).

The advantage of direct techniques is that results can be visualised directly after *in situ* hybridisation and that potential background staining by non-specific binding of immunocytochemical reagents does not occur. In contrast with indirect techniques, where the number of reporter molecules can be increased immunocytochemically, they can, however, be limited in their sensitivity. Obviously, when antibodies against the reporter molecule are available (e.g. anti-FITC, Bauman *et al.*, 1981), the favourable properties of both the direct and indirect techniques can be exploited with one nucleic acid labelling procedure.

Currently some six different nucleic acid hapten modification procedures are available. Some of them are based on (photo)chemical modification of the probe (e.g. AAF, Hg and sulfonation/transamination), others on enzymic incorporation of haptenised (deoxy)nucleotide triphosphates (e.g. biotin–dUTP and digoxigenin–dUTP). The number of different haptens to be used for non-isotopic hybridisation studies is likely to increase further and is already well above the number of spectrally well-separated immunofluorophores that are available at this moment. Possibly with the exception of the mercurated probes (Brown & Balmain, 1979) all haptenisation methods for nucleic acids have in common that they affect the hybridisation properties of the probe (i.e. rate of hybridisation and hybrid stability). For fairly large probes and for the degrees of modification generally used (1–10%), this effect is quite acceptable as it does not lead to a loss of hybridisation specificity in most applications. For modification of synthetic oligonucleotides, however, other routes of modification are recommended. For instance, covalently linking alkylamino groups to the 5' terminus of synthetic oligonucleotides may be used to couple any alkylamine reactive hapten (Agrawal *et al.*, 1986). Another approach for oligonucleotide modification would be the use of haptenised dXTPS in terminal transferase reactions (Riley *et al.*, 1986).

Another important feature of nucleic acid haptenisation procedures is the fact that the hapten should be accessible and recognisable in the duplex form for the anti-hapten antibody. When, for instance, the antibody is raised against a hapten-modified mononucleotide, the resulting antibodies may not all be suitable in hybridisation studies. As an example, antibodies raised against bromodeoxyuridine coupled to protein carriers will not recognise the BrdU in duplex BrdU-substituted DNA, whereas they are excellent with single-stranded DNA (Gonchoroff *et al.*, 1987).

The *in situ* hybridisation procedure

An *in situ* hybridisation procedure is a multi-step procedure and the final result is dependent on the quality of each of the individual steps. For some of the steps one can perform independent quality control. For instance, probe modification and immunological detection systems, can be conveniently tested by performing filter hybridisations (Leary *et al.*, 1983). The factor that is most critical for the success of an *in situ* hybridisation is the accessibility of the target sequences for the high molecular weight reagents such as DNA probes (strept) avidin and immunoglobulins (Raap *et al.*, 1989). Therefore, after primary fixation and microscope slide preparation one may have to perform pretreatments which increase accessibility. When necessary, this should be done in such a way that an optimum balance is found between the *in situ* hybridisation signal and preservation of morphology. Typical examples are proteinase K digestion (e.g. for chromosomes and cells), pepsin digestion (e.g. for paraffin-embedded formalin fixed tissues), RNase, freeze-thawing and detergent treatments.

For *in situ* DNA detection, target DNA denaturation has to be performed. There seems to be general consensus that heat denaturation with or without probe already present yields good results. Other significant points are the use of formamide in hybridisation solutions to reduce optimal hybridisation temperatures, and the fragmentation of probe molecules so as to promote probe penetration. In this respect synthetic oligonucleotides can be very useful. Their efficient *in situ* hybridisation, which is probably due to their good penetration properties, may in fact compensate to a large extent for the smaller target they cover (Dirks *et al.*, in preparation). Of special importance concerning the hybridisation is the fact that the labelled probe used in *in situ* hybridisation need not be absolutely pure. Probe sequences which may hybridise to other sequences than the target sequences can effectively be competed out by adding the relevant unlabelled competitor DNA (e.g. total human DNA). This approach has contributed significantly to rapid fine mapping of cosmid clones, yeast artificial chromosome clones, as well as to the use of whole chromosome-specific libraries as probes for so-called 'chromosome painting' (Landegent *et al.*, 1987; Cremer *et al.*, 1988; Lichter *et al.*, 1988; Pinkel *et al.*, 1988).

After washing the preparations at the desired stringency, immunocytochemical detection is performed. For non-biotin probes, indirect immunocytochemical techniques are most commonly used. For biotin-probes, a simple one-step procedure is performed. When high sensitivity is required, immunological amplification procedures can be performed. The procedure described by Pinkel *et al.* (1986) using biotinated anti-avidin and avidin–FITC is very useful in this respect, especially when combined with the high salt washes recommended by Lawrence *et al.* (1988). For fluorescence detection, the addition of anti-fading agents is strongly recommended.

For the major applications, operational protocols have emerged during the last few years and the reader is referred to the references.

Fluorescent reporter molecules

Three types of reporter molecules are widely used in immunocytochemistry and non-radioactive *in situ* hybridisation: fluorochromes, enzymes and colloidal gold. The choice as to what type of microscopy is going to be best in a particular experiment depends on several factors. The choice of the reporter molecule also determines the type of microscopy that is going to be used. In this chapter we will only discuss fluorescence microscopy. For many applications there can be very valid reasons to use other types of microscopy.

Detection sensitivity is an issue of constant concern in immunocytochemistry and hybridiocytochemistry. As an example, the *in situ* detection of a unique DNA sequence of 3 kb in a human genome requires the detection of as little as 3 attograms of DNA. Therefore, next to optimal *in situ* hybridisation procedures, optimal microscopy should also be employed. For fluorescence microscopy this involves the use of high numerical apertures, low magnification non-compound lenses (as far as compatible with high numerical aperture), high-energy light sources of proper spectral characteristics, high-quality filter sets and dichroic mirrors and proper adjustments of illumination (see Ploem & Tanke, 1987).

As it is of increasing importance to detect more than one target sequence simultaneously (e.g. for relative mapping of DNA sequences in meta and interphase nuclei, for detection of numerical and structural chromosome aberrations, for detection of multiple mRNA species in one cell), the fluorescence microscope should preferably be equipped with three filter sets allowing near UV, blue and green excitation. This allows for the use of three immunofluorophores emitting in the blue (aminomethyl coumarin acetic acid (AMCA), green (FITC) and red (TRITC, XRITC, Texas Red), respectively. The blue immunofluorophore AMCA has been developed recently by Robinson and coworkers (Khalfan et al., 1986) and its introduction allowed us for the first time to do triple fluorescence *in situ* hybridisation (Nederlof et al., 1989).

Recently, immunofluorophores based on cyanins, which emit in the infra red, have also been described (Waggonner, personal communication). Thus, provided no cross-reactivity occurs during immunocytochemical reactions it is feasible to perform four-colour fluorescence *in situ* hybridisation. An infrared sensitive detection device will, however, be necessary to visualise the infra-red emission. An elegant solution to the problem of the limiting number of fluorophores is the use of double haptenised probes. For example, we have used alphoid DNA probes for four different chromosomes labelled with biotin, AAF, sulphon and biotin plus AAF and detected the various probes with AMCA, FITC and TRITC, the fourth hybridisation site being identified by double fluorescence colours (Nederlof et al., in preparation). Obviously, this approach cannot be followed when two targets colocalise. In preliminary experiments we have performed microfluorometry on such double-labelled probes and have found that the ratio of the fluorescence intensities is fairly constant (Nederlof et al., in preparation). Thus, with fluorescence ratio imaging and with probes labelled with different hapten ratios, it is in principle possible to perform multiple hybridisation with only two haptens. In fact, this principle may form the basis of an hybridisation-based chromosome banding procedure. A new largely unexplored area in immunocytochemistry is the use of time-resolvable luminiscent reporter molecules. When combined with time-resolved luminescence microscopy, autofluorescence, one of the main disadvantages of fluorescence microscopy, can in principle be eliminated (Tanke et al., 1988).

Applications

Major applications of *in situ* hybridisation are found in (cancer) cytogenetics, cell biology, virology, and oncology. Chromosomal localisation of DNA sequences, both in inter- and metaphase by *in situ* hybridisation has become increasingly important in cytogenetics. The sensitivity of current protocols is such that single copy gene sequences of a few kb can be localised (Albertson *et al.*, 1985; Ambros *et al.*, 1986; Bhatt *et al.*, 1988; Garson *et al.*, 1987; Hopman *et al.*, 1986; Landegent, 1985; Lawrence *et al.*, 1988). Given this sensitivity, it is quite feasible to detect more repetitive sequences. Therefore the technique is applied amongst others for detection of species-specific DNA in somatic cell hybrids using simply total DNAs as probes, gene amplification, and transfected and transgenic DNA (Durnam *et al.*, 1985; Hopman *et al.*, 1986; Mitchell *et al.*, 1986; Schardin *et al.*, 1985). Of interest for solid tumour cancer cytogenetics, where it can be difficult to obtain sufficient good-quality metaphase chromosomes for Giemsa banding, is the use of chromosome specific satellite sequences, as their application allows for the assessment of numerical, and to some extent structural chromosome aberrations in interphase nuclei and low-quality metaphase spreads (Devilee *et al.*, 1988; Hopman *et al.*, 1988; Nederlof *et al.*, 1989). Here the multiple fluorescence hybridisations with chromosome (segment) specific labelling are of great help, since they will give more information concerning chromosomal heterogeneity in tumour cell nuclei, than the separate individual hybridisations (Hopman *et al.*, 1988; Nederlof *et al.*, 1989). These multiple hybridisations, especially when combined with confocal laser scanning microscopy and 3D reconstructions, are also very useful for documenting the structural organisation of the chromatin during cell differentiation in health and disease (Manuelidis & Borden 1988; Appels, 1989).

In diagnostic virology the *in situ* hybridisation methods have rapidly acquired a routine application, especially for those viruses that are difficult to diagnose by other methods. Typical examples in histopathology are cytomegalovirus, human papillomaviruses, JC virus, parvo virus B19 (Beckmann *et al.*, 1985; Brigati *et al.*, 1982; Burns *et al.*, 1986; Boerman *et al.*; 1989, Jiwa *et al.*, 1989; Myerson *et al.*, 1984a, 1984b; Raap *et al.*, 1988; Salimans *et al.*, 1989; Porter *et al.*, 1988; Walboomers *et al.*, 1988). Finally, the methods specifically aiming at microscopic mRNA localisation are applicable in various fields. When compared with DNA detection, there is, however, still a limited number of publications describing *in situ* RNA detection. It appears that careful analysis of the various steps of the procedure is essential (Lawrence & Singer, 1985; Dirks *et al.*, 1989). Using a biological model system which abundantly expresses specific mRNAs (neuropeptidergic systems of the pond snail *Lymnaea stagnalis*) we have successfully localised several mRNAs coding for neuropeptides using nonisotopic methods (Dirks *et al.*, 1989), including double hybridisation with synthetic oligonucleotides.

References

Agrawal, S., Christodoulow, C. & Gait, J. C. (1986). Efficient methods for attaching non-radioactive labels to the 5' end of synthetic oligodeoxyribonucleotides. *Nucleic Acids Research*, **14**, 6227–45.

Albertson, D. G. (1985). Mapping muscle protein genes by *in situ* hybridization using biotin-labelled probes. *The EMBO Journal*, **4**, 2493–8.

Ambros, P. F., Matzke, M. A. & Matzke, A. J. M. (1986). Detection of a 17 kb unique (T-DNA) in plant chromosomes by *in situ* hybridization. *Chromosoma*, **94**, 11–16.

Appels, R. (1989). Three dimensional arrangement of chromatin and chromosomes: old concepts and new techniques. *Journal of Cell Science*, **92**, 325–8.

Bauman, J. G. J., Wiegant, J., Borst, P. & van Duijn, P. (1980). A new method for fluorescence microscopical localization of specific DNA sequences by *in situ* hybridization of fluorochrome labeled RNA. *Experimental Cell Research*, **138**, 485–90.

Bauman, J. G. J., Wiegent, J. & van Duijn, P. (1981). Cytochemical hybridization with fluorochrome-labelled RNA. III. Increased sensitivity by the use of anti-fluorescein antibodies. *Histochemistry*, **73**, 181–93.

Beckmann, A. M., Myerson, D., Daling, J. R., Kivat, N. B., Fenoglio, C. M. & McDougall, J. K. (1985). Detection and localization of human papillomavirus in human genital condylomas by *in situ* hybridization with biotinylated probes. *Journal of Medical Virology*, **16**, 165–73.

Bhatt, B., Burns, J., Flannery, D. & McGee, J. O. D. (1988). Direct visualisation of single copy genes on banded chromosomes by nonisotopic *in situ* hybridization. *Nucleic Acids Research*, **16**, 3951–61.

Boerman, R. H., Arnoldus, E. P. J., Raap, A. K., Peters, A. C. B., Ter Schegget, J. & van der Ploeg, M. (1989). Diagnosis of progressive multifocal leukoencephalopathy by hybridization techniques, *Journal of Clinical Pathology*, **42**, 153–61.

Boom, R., Geelen, J. L., Sol, C. J., Raap, A. K., Minnaar, R. P., Klaver, B. P. & Noordaa van der, J. (1986). Establishment of a rat cell line inducible for the expression of human cytomegalovirus immediate early gene products by protein synthesis inhibition. *Journal of Virology*, **58**, 851–9.

Brigati, D. J., Myerson, D., Leary, J. J., Spalholz, B., Travis, S., Fong, C. K. Y., Hsiung, G. D. & Ward, D. C. (1982). Detection of viral genomes in cultured cells and paraffin-embedded tissues sections using biotin-labelled hybridization probes. *Virology*, **126**, 32–50.

Brown, T. D. K. & Balmain, A. (1979). The effects of mercury-substitution on the hybridization characteristics of nucleic acids. *Nucleic Acids*

Research, **7**, 2357–68.
Burns, J., Redfern, D. R. M., Esiri, M. M. & McGee J. O'D. (1986). Human and viral gene detection in routine paraffin embedded tissue by *in situ* hybridization with biotinylated probes: viral localisation in herpes encephalitis. *Journal of Clinical Pathology*, **39**, 1066–73.
Cremer, T., Lichter, P., Borden, J., Ward, D. C. & Manuelidis, L. (1988). Detection of chromosome aberrations in metaphase and interphase tumor cells by *in situ* hybridization using chromosome specific library probes. *Human Genetics*, **80**, 235–46.
Devilee, P., Thierry, F., Kievits, T., Kolluri, R., Hopman, A. H. N., Willard, H. F., Pearson, P. L. & Cornelisse, C. J. (1988). Detection of chromosome aneuploidy in interphase nuclei from human primary breast tumors using chromosome specific repetitive DNA probes. *Cancer Research*, **48**, 5825–30.
Dirks, R. W., Raap, A. K., van Minnen, J., Vreugdenhil, E., Smit, A. B. & van der Ploeg, M. (1989). Detection of mRNA, molecules coding for neuropeptide hormones of the pond snail *Lymnaea stagnalis* by radioactive and non-radioactive *in situ* hybridization: a model study for mRNA detection. *Journal of Histochemical Cytochemistry*, **37**, 7–14.
Durnam, D. M., Gelinas, R. E. & Myerson, D. (1985). Detection of species chromosomes in somatic cell hybrids. *Somatic Cell Molecular Genetics*, **11**, 571–7.
Garson, J. A., van den Berghe, J. A. & Kemshead, J. T. (1987). Novel non-isotopic *in situ* hybridization technique detects small (1 kb) unique sequences ion routinely G-banded human chromosomes: fine mapping of N-*myc* and b-*NGF* genes. *Nucleic Acid Research*, **15**, 4761–70.
Gonchoroff, N. J., Katzmann, J. A., Currie, R. M., Evans, E. L., Houck, D. W., Kline, B. C., Greipp, P. R. & Loken, M. R. (1987). S-phase detection with antibody to bromodeoxyuridine. Role of DNase pretreatment. *Journal of Immunology Methods*, **xx**, 97–101.
Hopman, A. H. N., Wiegant, J., Tesser, G. I. & van Duijn, P. (1986a). A non-radioactive *in situ* hybridization method based on mercurated nucleic acid probes and sulfhydryl–hapten ligands. *Nucleic Acids Research*, **14**, 6471–88.
Hopman, A. H. N., Wiegant, J., Raap, A. K., Landegent, J. E., van der Ploeg, M. & van Duijn, P. (1986b). Bi-color detection of two target DNAs by non-radioactive *in situ* hybridization. *Histochemistry*, **85**, 1–4.
Hopman, A. H. N., Ramaekers, F. C. S., Raap, A. K., Beck, J. L. M., Devilee, P., van der Ploeg, M., & Vooij, G. P. (1988). *In situ* hybridization as a tool to study numerical chromosome aberrations in solid bladder tumors. *Histochemistry*, **89**, 307–16.
Jiwa, N. M., Raap, A. K., van de Rijke, F. M., Mulder, A., Weening, J., Zwaan, F. E., The, T. H. & van der Ploeg, M. (1989). Detection of human cytomegalovirus antigens in formaldehyde fixed material from bone marrow transplant patients. *Journal of Clinical Pathology*, in press.
John, H., Birnstiel, M. & Jones K. (1969). RNA–DNA hybrids at the cytological level. *Nature (London)*, **223**, 582–7.

Khalfan, H., Abuknesha, R., Rand-Weaver, R., Price, R. G. & Robinson, D. (1986). Aminomethyl coumarin acetic acid: a new fluorescent labeling agent for proteins. *Histochemical Journal*, **18**, 497–9.

Landegent, J. E., Jansen in de Wal, N., Ommen, G. J. B., Baas, F., De Vijlder, J. J. M., van Duijn, P. & van der Ploeg, M. (1985). Chromosomal localization of a unique gene by non-autoradiographic *in situ* hybridization. *Nature (London)*, **317**, 175–7.

Landegent, J. E., Jansen in de Wal, N., Fisser-Groen, Y. M., Bakker, E., van de Ploeg, M. & Pearson, P. L. (1986). Fine mapping of the Huntingdon disease linked D4S10 locus by non-radioactive *in situ* hybridization. *Human Genetics*, **73**, 354–7.

Landegent, J. E., Jansen in de Wal, N., Dirks, R. W., Baas, F. & van de Ploeg, M. (1987). Use of whole cosmid cloned genomic sequences for chromosomal localization by non-radioactive *in situ* hybridization. *Human Genetics*, **77**, 366–70.

Lawrence, J. B. & Singer, R. H. (1985). Quantitative analysis of *in situ* hybridization methods for the detection of actin gene expression. *Nucleic Acids Research*, **13**, 1777–99.

Lawrence, J. B., Villnave, C. A. & Singer, R. H. (1988). Sensitive high resolution chromatin and chromosome mapping *in situ*: presence and orientation of two closely integrated copies of EBV in a lymphoma line. *Cell*, **52**, 51–61.

Leary, J. L., Brigati, D. J. & Ward, D. C. (1983). Rapid and sensitive colorimetric method for visualizing biotin-labeled DNA probes hybridized to DNA or RNA immobilized on nitrocellulose: Bio-blots. *Proceedings of the National Academy of Sciences, USA*, **80**, 4045–9.

Lichter, P., Cremer, T., Borden, J., Manuelidid, L. & Ward, D. C. (1988). Delination of individual human chromosomes in metaphase and interphase cells by *in situ* hybridization using recombinant DNA libraries. *Human Genetics*, **80**, 224–34.

Manuelidis, L. & Borden, J. (1988). Reproducible compartimentalization of individual chromosome domains in human CNS cells revealed by *in situ* hybridization and three dimensional reconstruction. *Chromosoma (Berl.)*, **96**, 397–410.

Mitchell, A. R., Ambros, P., Gosden, J. R. & Porteous, D. J. (1986). Gene-mapping and physical arrangements of human chromatin in transformed, hybrid-cells-fluorescent and autoradiographic *in situ* hybridization compared. *Somatic Cell Molecular Genetics*, **12**, 313–24.

Myerson, D., Hackman, R. C., Nelson, J. A., Ward, D. C. & McDougall, J. K. (1984a). Widespread occurence of histologically occult cytomegalovirus. *Human Pathology*, **15**, 430–9.

Myerson, D., Hackman, R. C. & Myers, J. D., (1984b). Diagnosis of cytomegalovirus pneumonia by *in situ* hybridization. *Journal of Infectious Diseases*, **150**, 272–7.

Nederlof, P. M., Robinson, D., Wiegent, J., Hopman, A. H. N., Tanke, H. J., Raap, A. K. (1989). Three color fluorescence for the simultaneous detection of multiple DNA targets. *Cytometry*, **10**, in press.

Pardue, M. L. & Gall, J. G. (1969). Molecular hybridization of radioactive DNA to the DNA of cytological preparations. *Proceedings of the National Academy of Sciences, USA*, **64**, 600–4.

Pinkel, D., Straume, T. & Gray, J. W. (1986). Cytogenetic analysis using quantitative, high sensitivity fluorescence hybridization. *Proceedings of the National Academy of Sciences, USA*, 2934–8.

Pinkel, D., Landegent, J., Collins, C., Fuscoe, J., Segraves, R., Lucas, J. & Gray, J. (1988). Fluorescence *in situ* hybridization with human chromosome specific libraries: detection of trisomy 21 and translocation of chromosome 4. *Proceedings of National Academy of Sciences, USA*, **85**, 9138–42.

Ploem, J. S. & Tanke, H. J. (1987). *Introduction to Fluorescence Microscopy*. RMS Microscopy Handbooks Series, no. 10: Oxford Science Publications.

Porter, H. J., Khong, T. Y., Evans, M. F., Chan, V. T. W. & Fleming, K. H., (1988). Parvovirus as a cause of hydrops fetalis. Detection by *in situ* hybridization. *Journal of Clinical Pathology*, **41**, 381–3.

Raap, A. K., Geelen, J. L., van der Meer, J. W. M., van den Boogaard, P. & van der Ploeg, M. (1988). Non-radioactive *in situ* hybridization for the detection of cytomegalovirus infections. *Histochemistry*, **88**, 367–73.

Raap, A. K., Hopman, A. H. N., van der Ploeg, M. (1989). Use of hapten modified nucleic acid probes in DNA *in situ* hybridization. In: *Techniques in Immunocytochemistry*. (eds Bullock, G. & Petrusz, P.) vol 4, New York: Academic Press.

Renz, M. & Kurz, C. (1984). A colorimetric method for DNA hybridization. *Nucleic Acids Research*, **12**, 3435–44.

Riley, L. K., Marshall, M. E. & Coleman, M. S. (1986). A method for biotinylating oligonucleotide probes for use in molecular hybridizations. *DNA*, **5**, 333–7.

Salimens, M. M. M., van de Rijke, F. M., Raap, A. K. & van Elsacker-Niele, A. M. W. (1989). Detection of parvo virus B19 DNA in fetal tissue by *in situ* hybridization and the polymerase chain reaction. *Journal of Clinical Pathology*, **42**, 525–30.

Schardin, M., Cremer, T., Hager, H. D. & Lang, M. (1985). Specific staining of human chromosomes in Chinese hamster × man hybrid cell lines demonstrates interphase chromosomes territories. *Human Genetics*, **71**, 281–7.

Wachtler, F., Hopman, A. H. N., Wiegant, J. & Schwarzacher, H. G. (1986). On the position of nucleolus organizer regions (NORs) in interphase nuclei. Studies with a new, non-autoradiographic *in situ* hybridization method, *Experimental Cell Research*, **167**, 227–40.

Walboomers, J. M. M., Mullink, H., Melchers, W. J. G., Meyer, C. J. L. M., Struyk, A., Quint, W. G. J., van der Noordaa & Ter Schegget, J. (1988). Sensitivity of *in situ* detection with biotinylated probes of HPV 16 DNA in frozen tissue sections of squamous cell carcinoma of the cervix. *American Journal of Pathology*, **131**, in press.

A. Giaid, S. J. Gibson, J. H. Steel, P. Facer and J. M. Polak

The use of complementary RNA probes for the identification and localisation of peptide messenger RNA in the diffuse neuroendocrine system

Introduction

The introduction of *in situ* hybridisation for the localisation of specific peptide mRNAs at the site of synthesis is a useful adjunct to immunocytochemistry and has been used here to provide some answers concerning the expression of peptide mRNAs in the diffuse neuroendocrine system. Complementary ribonucleotide probes were transcribed from subcloned inserts of cDNAs encoding specific peptide sequences. The peptides studied include the neuropeptides calcitonin gene-related peptide (CGRP), substance P, vasoactive intestinal polypeptide (VIP) and neuropeptide Y (NPY) as examples of neurally borne substances in the central and peripheral nervous system and atrial natriuretic peptide (ANP) and prolactin as examples of circulating hormones produced by endocrine cells. These examples emphasise the value of *in situ* hybridisation by elucidating sites of peptide mRNA production in the neuroendocrine system and their change in different physiological conditions.

It is now widely accepted that regulatory peptides are present in the typical endocrine cells of the 'diffuse endocrine system' described by Feyrter (1938) or the APUD (*a*mine or amine *p*recursor *u*ptake and *d*ecarboxylation) system of Pearse (1969) as well as in nerves of the central and peripheral nervous system. This newly recognised 'diffuse neuroendocrine system' is thus characterised by the synthesis and release of a number of regulatory peptides which mediate the actions of these cells in a number of varied ways. Some act as circulating hormones, while others have a more restricted effect,

Abbreviations
ANP, atrial natriuretic peptide; ATP, adenosine triphosphate; cDNA, complementary deoxyribonucleic acid; CGRP, calcitonin gene-related peptide; cRNA, complementary ribonucleic acid; CTP, cytidine triphosphate; DEPC, diethylpyrocarbonate; DNA, deoxyribonucleic acid; GTP, guanidine triphosphate; H, tritium; mRNA, messenger ribonuclei acid; NPY, neuropeptide Y; P, phosphorus; RNA, ribonucleic acid; S, sulphur; SSC, standard sodium citrate; TCA, trichloroacetic acid; UTP, uridine triphosphate; VIP, vasoactive intestinal polypeptide.

acting only in the region of their release from endocrine or neuronal cells as potent modulators, neurotransmitters or long-term growth factors.

Although it was known for some years that peptides such as oxytocin and vasopressin (Du Vigneaud, 1956) were present in neurones of the central nervous system, it was the discovery that the undecapeptide, substance P was present, not only in some enterochromaffin cells but also in nerves of the gastrointestinal mucosa (Pearse & Polak, 1975; Nilsson et al., 1975) and central nervous system (Hökfelt et al., 1975) that provided a foundation for the relationship between the diffuse endocrine and nervous systems. Since then, many peptides have been localised to both endocrine and neural systems, and include calcitonin gene-related peptide, the enkephalins, vasoactive intestinal polypeptide, somatostatin and atrial natriuretic polypeptide, to name only a few.

Immunocytochemistry has been invaluable in the delineation and analysis of peptide-immunoreactive cell types in the diffuse neuroendocrine system in both health and disease. However, a major drawback to this technique is that the localisation of an intracellular peptide/protein antigen tells little about the intracellular processes concerned with protein synthesis. Advances in molecular biology have provided a new dimension to the study of the neuroendocrine system by allowing investigation of peptide gene expression within histological preparations. In particular, the *in situ* hybridisation method allows the localisation, at the cellular level, of messenger RNA (mRNA) species directing the synthesis of peptides. The recent introduction of this method, which is based on the ability of complementary sequences of nucleic acids to hybridise with the cytoplasmic constituents of the cell, has therefore provided morphologists with a valuable tool for the investigation of peptide/protein biosynthesis and has added a new dimension to our interpretation of cellular events. The technique of *in situ* hybridisation was first described by Gall and Pardue in 1969 (Gall & Pardue, 1969). Since then, a variety of nucleic acid probes have been used for *in situ* hybridisation. Double-stranded DNA (Pardue, 1985) and single-stranded DNA (Stevens et al., 1987) or complementary RNA probes and oligonucleotides (Cox et al., 1984) can be constructed, which are complementary to the intracellular mRNA. The precise nature of the complementary base pairing between nucleic acid probes and their corresponding mRNAs within the tissue allows detection, *in situ*, of both high- and low-abundance mRNAs.

In order to visualise the resultant DNA–RNA or RNA–RNA hybrid in the tissue under investigation, the nucleic acid probe must first be labelled. The choice of label falls into two categories (Fig. 1). Radiolabelled probes using ^{32}P, ^{35}S or ^{3}H are most commonly employed. Alternatively, they can be labelled with non-radioactive tags, the most popular of which uses biotinylated nucleotides which are indirectly detected via avidin whose signal

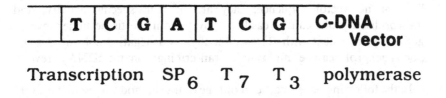

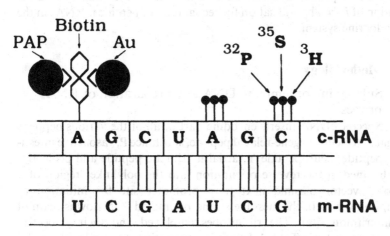

Figure 1. A schematic representation of the *in situ* hybridisation method using a cRNA probe. The cRNA probe, generated by transcription of the cDNA template, shows some of the possible labelling methods which are available to visualise the RNA–RNA hybrid.

can be amplified with either a light visible or a fluorescent molecule. However, non-isotopic methods are of relatively poor sensitivity compared to radioactive methods and thus are not recommended for detection of peptide mRNAs of low copy numbers (Singer, Lawrence & Villnave, 1986). The various types of nucleic acid probe have been described in detail elsewhere (Chapters 1 and 2). For the majority of our studies on peptide gene expression in the diffuse neuroendocrine system, we have utilised radiolabelled, complementary RNA (cRNA) probes (but see Denny *et al.*, 1988; Hamid *et al.*, 1988).

The use of RNA probes offers several advantages over DNA oligonucleotide probes (Cox *et al.*, 1984). Firstly, and most importantly, cRNA probes usually provide higher specific signals than DNA probes, since there is no strand in the hybridisation solution (i.e. mRNA) to compete with mRNAs in the tissue for annealing with the labelled cRNA probe in the solution. In addition, the stability of the RNA–RNA hybrid is greater than that of RNA–DNA hybrids. Furthermore, RNA probes are of fixed length and contain little or no vector sequence, they are of high specific activity, and unhybridised probe can be easily removed from the tissue under investigation by treatment with RNase. Finally, as a negative control, a sense (message) probe can be employed by transcribing from the cDNA in reverse orientation.

In the following pages a general outline of the method using radioactively labelled cRNA probes is described and includes some examples of the application of *in situ* hybridisation for recognition of peptide mRNAs in the neuroendocrine system.

Methodology

Subcloning of template DNA and preparation of RNA probes

Specific cDNA inserts encoding part or all of the various peptides investigated (calcitonin gene-related peptide, substance P, vasoactive intestinal polypeptide, neuropeptide Y, atrial natriuretic peptide and prolactin) were subcloned in the reverse orientation into the polylinker region of a 'riboprobe' vector plasmid next to a promoter site for initiation of transcription (i.e. with the message strand orientated 3'-5' downstream of the transcription site). The riboprobes employed and their respective promoters are shown in Table 1. Prior to transcription the circular plasmid is linearised by restriction endonuclease digestion at the 3' end of the insert or at a suitable site within the vector (see Table 1 for restriction enzymes). Riboprobes, complementary to their specific peptide coding sequences, were generated from the linearised templates using an appropriate RNA polymerase in the presence of radioactively labelled ribonucleotides (page 60). Probes of 300–500 bases generally give good tissue penetration and reliable results. Probes longer than this were shortened by alkaline hydrolysis prior to hybridisation (page 61).

Tissue preparation and processing

Rat and human tissues were obtained as fresh as possible. Human tissues were fixed for *in situ* hybridisation by immersion in 4% paraformaldehyde in 0.1 M borate buffer (pH 9.5) for 4–6 h. Rats were anaesthetised and fixed by intracardial perfusion with the above mentioned fixative. Tissues were dissected and immersed in a fresh solution of fixative for a further 2–3 h (pages 61–2). Tissues were then washed in 0.1 M phosphate-buffered saline containing 15% sucrose and stored overnight at 4°C.

Frozen tissue blocks were made and tissue sections (8–15 μm) were prepared using a cryostat and mounted on poly-L-lysine coated slides (Huang *et al.*, 1983) which had been previously cleaned and baked (150°C)

Table 1. *Summary of riboprobes systems employed to generate cRNA probes*

Peptide cDNA inserts	Riboprobe vector	Promoter	Restriction enzymes	Source
rat α-CGRP	pSP64	SP6	EcoRI	Amara, USA
rat β-CGRP	pSP64	SP6	EcoRI	Amara, USA
human α-CGRP	pSP64	SP6	EcoRI	Craig, UK
rat β-PPT	pGEM1	SP6	Hind III	Krause, USA
human β-PPT	GEM 3	SP6	AVA	Harmar, UK
rat VIP	pSP64	SP6	pVu 11	Goodman, USA
human VIP	pSP65	SP6	Bam H1	Okamoto, Japan
rat NPY	pSP64	SP6	EcoRI	Dixon, USA
human NPY	pSP62-PL	SP6	Bam HI	Dixon, USA
rat ANP	PSP64	SP6	EcoRI	Nakazato, Japan
human ANP	PSP64	SP6	Sst/EcoRI	Nakazato, Japan
rat prolactin	PSP65	SP6	EcoRI	Maurer, USA

CGRP, calcitonin gene-related peptide; β-PPT, β preprotachykinin; VIP, vasoactive intestinal polypeptide; NPY, neuropeptide Y; ANP, atrial natriuretic peptide.

overnight, or alternatively, treated with a 2% solution of Absolve (Dupont, UK Ltd, Herts, UK) for 2 h to remove any trace of RNase. The slides were dried (37°C) overnight before processing for hybridisation.

Pre-hybridisation

Pre-hybridisation treatment serves two purposes. Firstly, treatment of the tissue sections with Triton X-100 and proteinase K, permeabilises and deproteinises the tissue thus ensuring easy penetration of the probe. Secondly, an acetylation step is included which acts to reduce electrostatic, non-specific binding of the probe to the section (page 62).

Hybridisation

The transcribed probe is diluted to give an approximate concentration of 5×10^5–5×10^6 cpm per section. The precise concentration used varies and will depend on the efficiency of the probe and the abundance of the mRNA investigated. The probe concentration should therefore be derived empirically. The tissue section, once covered with dilute probe, is allowed to incubate for 12–18 h in a humid atmosphere (page 62). The stringency of the hybridisation reaction can be increased to within 25°C of the melting temperature (T_m) of the RNA–RNA duplex investigated (i.e. 25°C + the temperature at which 50% of the hybrids dissociate).

Post-hybridisation washing

Post-hybridisation treatments are employed to reduce background levels in the tissue sections. Unbound probe is removed by a series of washes in decreasing concentrations of standard sodium citrate and by treatment with RNase (page 63).

Autoradiography

Tissue sections can be dipped in liquid photographic emulsion (page 63), or for a rapid appraisal (1 day exposure time) can be apposed directly to X-ray film and exposed at room temperature or at -80°C with an intensifier screen and, once the result is evaluated, can be subsequently dipped. The dipping method provides a more precise technique for visualisation of the label at the cellular level whereas the film method affords a gross, low-power view. The latter method is thus of particular use when the mRNA is concentrated within an anatomically defined region (e.g. prolactin in the anterior lobe of the pituitary). The exposure time for emulsion-coated slides is variable and dependent on the particular radiolabel and abundance of message. Generally, ^{35}S will give a slower result (1–17 days) than ^{32}P (5–10 days).

Controls

To assess the specificity of the signal obtained the following controls are routinely employed.

1. The use of probe identical to the coding strand (sense probe).
2. Pre-hybridisation treatment of sections with RNase A (20 µg/ml 37°C, 30 min) to destroy all mRNA.
3. The use of an inappropriate cRNA probe or an inappropriate tissue.
4. If available, the use of a cRNA probe from a different segment of the same genome.

In addition, Northern blot analysis (for estimation of total cellular mRNA) of microdissected regions and comparison of *in situ* hybridisation sections with serial sections which have been immunostained for the corresponding peptide are good indications of the specificity of the hybridisation result.

Expression of peptide mRNAs in the neuroendocrine system

Calcitonin gene-related peptide (CGRP)

CGRP is a 37 amino acid peptide produced by alternative processing of the primary transcript of the calcitonin gene (Amara *et al.*, 1982) (Fig. 2). Two forms of CGRP have been identified in both rat (Amara *et al.*, 1985) and man (Morris *et al.*, 1984) and are derived from related α and β genes (Fig. 3). Immunocytochemical studies have shown that CGRP, although present in some endocrine cells, is predominantly expressed in neural tissues. In particular, CGRP has been localised to numerous sensory neurones, consistent with several lines of evidence suggesting that CGRP may participate in the mediation of nociception. However, CGRP immunoreactivity has also been localised to motoneurones (Rosenfeld *et al.*, 1983) (Gibson *et al.*, 1984) and to motor end-plates in a number of species including man (Takami *et al.*, 1985; Matteoli *et al.*, 1988). The granular appearance of CGRP immunostaining associated with motoneurones has led, however, to speculation that the immunoproduct is due to CGRP-immunoreactive fibres synapsing on the motoneural membrane rather than to *de novo* synthesis of the peptide. Thus, the technique of *in situ* hybridisation was employed to determine whether CGRP is an authentic product of motoneurones.

Tissue sections from human and rat spinal cord and (as positive controls) dorsal root ganglia were hybridised with human or rat α and β CGRP probes respectively (Gibson *et al.*, 1988). In both species, the signals obtained using α CGRP probes were stronger than for the β probe. In the dorsal root ganglia,

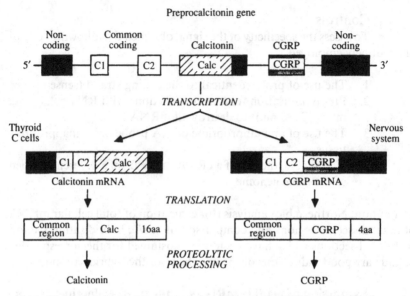

Figure 2. Schematic representation of alternative RNA splicing pathways in the expression of the preprocalcitonin gene. The preprocalcitonin gene shown encodes calcitonin (CALC) and α CGRP. aa, amino acids.

Primary sequence of rat and human α and β CGRP

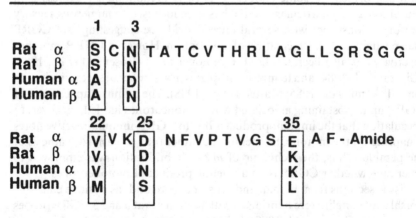

Figure 3. The amino acid sequences of α and β CGRP in rat and man. β-CGRP differs from the previously identified α form by one amino acid in rat and three in man.

numerous sensory neurones of small and medium sizes were labelled with silver grains (Fig. 4(a), (b)). In the ventral horn of the spinal cord, most large neurones (presumptive motoneurones) were similarly labelled (Fig. 4(c)). Comparison with adjacent sections, immunostained with antisera to CGRP showed a close parallel between the numbers of neurones positively labelled by *in situ* hybridisation and immunocytochemistry in the dorsal root ganglia and rat spinal cord. In human spinal cord, while most motoneurones expressed CGRP mRNA, only a few motoneurones were immunostained, which suggested that the mature peptide is more rapidly degraded than the message in post-mortem tissues. Thus, *in situ* hybridisation has shown unequivocally that motoneurones are able to synthesise CGRP. Since a muscle trophic rôle has recently been proposed for this peptide (Fontaine *et al.*, 1986; New & Mudge, 1986) its potential importance in the growth and maintenance of motor nerves should not be overlooked.

Although α CGRP mRNA was found to predominate in the spinal cord and dorsal root ganglia, it was of interest to see whether any class of neurones showed a differential expression of these genes. A likely region to investigate this was the gastrointestinal tract. Immunocytochemistry has shown that following treatment with a sensory neurotoxin, capsaicin, many CGRP-immunoreactive nerves disappear from the stomach but not the colon. Thus, the source of the fibres in the colon is thought to be from intrinsic, enteric neurones which can be visualised immunocytochemically following colchicine treatment. Using α and β CGRP probes it was possible to demonstrate that enteric neurones in the myenteric plexus expressed β CGRP (Fig. 4(d)) but not α CGRP transcripts (Mulderry *et al.*, 1988).

Substance P

Substance P is an 11 amino acid peptide and the best-known member of the tachykinin family. Two precursor mRNAs have been isolated, one encoding the substance P sequence (α-preprotachykinin) and the other encoding substance P and neurokinin A sequences (β-preprotachykinin) (Nawa *et al.*, 1983). There appears to be only one gene encoding for both preprotachykinins, with alternative splicing resulting in either mRNA (Nawa *et al.*, 1984) (Fig. 5).

Substance P has an extensive distribution throughout the body, it is present in most peripheral tissues and has a widespread distribution in the central nervous system (for review see Jordan & Oehme, 1985). However, the immunocytochemical localisation of substance P to sensory neurones, together with pharmacological and physiological data has implicated this peptide as a sensory neurotransmitter (Henry, 1977; Henry, 1982; Salt & Hill, 1983).

In situ hybridisation was performed using a cRNA encoding part of the β-

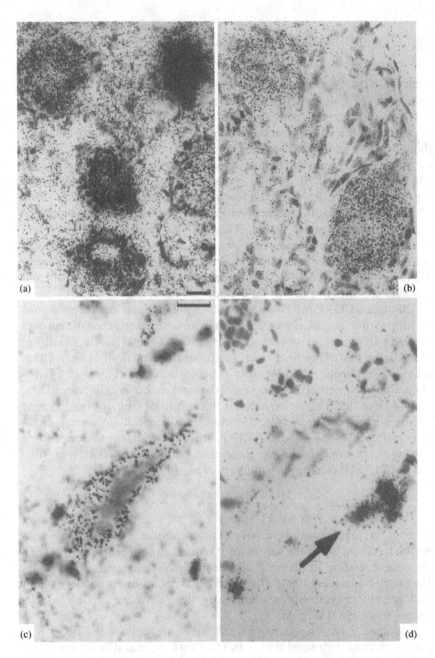

Figure 4. Autoradiograms of tissue sections hybridised with ^{32}P-labelled α-CGRP cRNA (a–c) and β-CGRP (d). Numerous sensory neurones are labelled in rat (a) and human (b) dorsal root ganglia. Scale bar = 25 μm. (c) One motoneurone from the lumbar ventral spinal cord of man. Silver grains depicting sites of α-CGRP mRNA production are seen overlying the cytoplasm. Scale bar = 25 μm. (d) Groups of enteric neurones (arrows) in the myenteric plexus of rat colon are labelled with silver grains following *in situ* hybridisation with a β-CGRP cRNA probe. (a)–(d) Counterstained with haematoxylin. Scale bar = 20 μm.

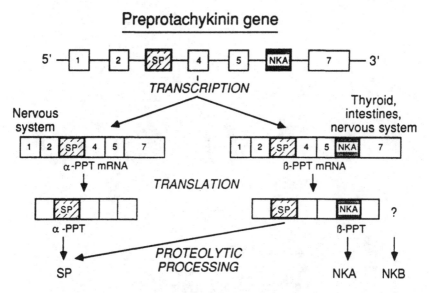

Figure 5. Schematic representation of alternative RNA splicing pathways in the expression of the preprotachykinin (PPT) gene. The PPT gene encodes both substance P (SP) and neurokinin A (NKA). A γ-PPT is also generated from the PPT gene through different RNA splicing events. Neurokinin B (NKB) is another member of the tachykinin family but the biosynthetic pathway for its generation is not shown.

preprotachykinin gene. Numerous labelled neurones were found in the brain, including the hypothalamus and amygdala (Denny et al., 1988) (Fig. 6(a)), in ganglion cells of the myenteric plexus of the colon (Denny et al., 1988; Facer et al., 1988) (Fig. 6(b)) and in small- and medium-sized neurones in dorsal root ganglia of rat and man (Fig. 6(c)). These localisations corresponded well with the immunocytochemical localisation of substance P.

Vasoactive intestinal polypeptide (VIP)

VIP is a 28 amino acid peptide which was isolated from porcine intestine by Said and Mutt in 1970 (Said & Mutt, 1970). VIP-immunoreactive neurones were first reported in the gastrointestinal tract and hypothalamus in 1976 (Larsson et al., 1976). Since this time, VIP has been found in most tissues of the body (Polak & Bloom, 1982). In the central nervous system, VIP has a selective distribution in the cerebral cortex and is present in other regions including the amygdala, periaqueductal grey and spinal cord. In peripheral nerves, most VIP immunoreactivity orginates from autonomic or enteric neurones (Said, 1984).

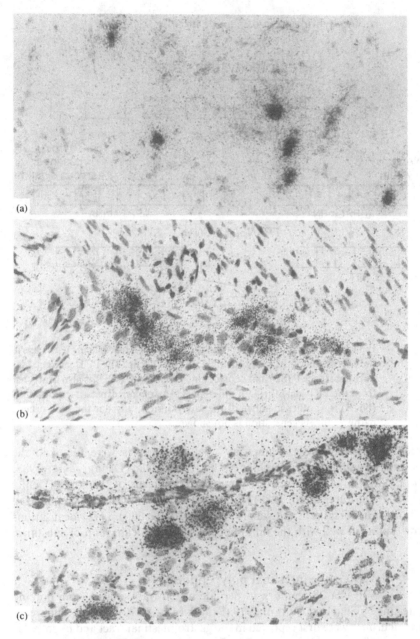

Figure 6. Autoradiograms of (a) rat brain (b) rat stomach and (c) human dorsal root ganglia following *in situ* hybridisation using a ^{32}P-labelled β-preprotachykinin cRNA probe. (a) Numerous neurones are labelled in the brain including the amygdaloid complex as shown here. (b) In the rat stomach silver grains are localised to neuronal cell bodies in the myenteric plexus. Scale bar = 20 μm. (c) In the dorsal root ganglia βPPT mRNA is localised to the small subpopulation of sensory neurones. Scale bar = 17 μm. (a)–(c) Counterstained with haematoxylin.

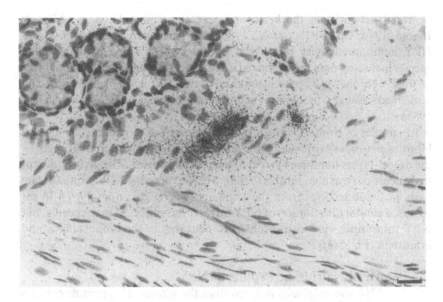

Figure 7. VIP mRNA localised by *in situ* hybridisation of ^{32}P-labelled VIP cRNA in rat colon. Silver grains overlie neuronal cell bodies in the submucous plexus. Counterstained with haematoxylin. Scale bar = 20 μm.

In human tissues, obtained after surgical resection or as soon as possible post-mortem, and in rat brain, *in situ* hybridisation using VIP cRNA probes (human or rat where appropriate) showed numerous VIP transcripts in the cerebral cortex and other brain regions. The distribution of cell bodies expressing VIP in the brain closely paralleled the known immunocytochemical localisation. In tissue sections of rat and human gut, hybridisation signals were observed in ganglion cells of the myenteric and submucous plexuses of both species (Fig. 7). However, comparison of hybridisation patterns obtained with the VIP probe and immunostained sections showed a greater number of cells producing VIP mRNA than could be detected by immunostaining for VIP (Facer *et al.*, 1988). Thus, *in situ* hybridisation may provide a more sensitive technique than immunocytochemistry. In addition, it provides a means to identify the potential site of peptide production, without the use of potent pharmacological agents such as colchicine. Such agents are often employed to increase stored peptide in neuronal cell bodies and hence aid immunocytochemical detection.

Neuropeptide Y (NPY)

This 36 amino acid peptide, was first isolated and sequenced from porcine brain (Tatemoto, 1982), and NPY has been localised to extensive neuronal networks in the brain and several peripheral tissues (Allen & Bloom, 1986; Chan-Palay et al., 1985a). In particular, the distribution of NPY immunoreactivity to cortical interneurones in several mammalian species including man has attracted attention. Although the functional rôle of these neurones is at present unknown, they display morphological and numerical changes in certain neurological diseases (Chan-Palay et al., 1985b).

In samples of frontal cortex, obtained at post-mortem and from surgical biopsies, neurones expressing NPY mRNA were found scattered throughout the deeper layers (laminae IV–VI) of the cerebral cortex. The distribution of these neurones was comparable to that of cells expressing NPY immunoreactivity in tissue sections from the same area (Terenghi et al., 1987). In rat tissues, a similar distribution of NPY mRNA was observed in the cortex, but NPY transcripts were also localised to neurones in the hippocampus and brainstem (Fig. 8(a)).

Atrial Natriuretic Peptide (ANP)

ANP was originally isolated from the atrium. This peptide consists of either 151 or 152 amino acids in human or rat heart respectively (Atlas et al., 1984; Kanagawa et al., 1984; Maki et al., 1984; Oikawa et al., 1984; Yamanaka et al., 1984). ANP possesses potent diuretic and natriuretic actions. Immunocytochemistry has shown that ANP is present in atrial myocytes in a number of animals including man (Kikuchi et al., 1987). However, in some lower vertebrates ANP-like immunoreactivity is found, not only in atria but also in the ventricles (Chapeau et al., 1985). These observations together with Northern blot analysis (Lattion et al., 1986) and radioimmunoassay of ventricular extracts (Gardener et al., 1986) indicated that the ventricle may also contribute to the production of ANP. To investigate whether the ANP gene was expressed in ventricular myocytes, in situ hybridisation was carried out on tissue sections from rat and human and cultures of rat heart using cRNA probes for rat and human ANP.

In tissue sections of both rat and man, many myocytes expressed ANP transcripts (Fig. 8(b)). In the ventricles however, labelled myocytes, although present, were fewer and were distributed mainly in the sub-endocardium of the intraventricular septum. Positively labelled cells were seen in cultures of rat atrium and ventricle, with greater numbers of cells found in the atrium. Comparison of the distribution of ANP mRNA in tissue sections and cultures showed a close parallel with the distribution of ANP immunoreactivity as shown by immunocytochemistry (Hamid et al., 1987).

Prolactin

The anterior pituitary consists of a mixture of endocrine cell types, all of which produce specific trophic hormones and whose activity is

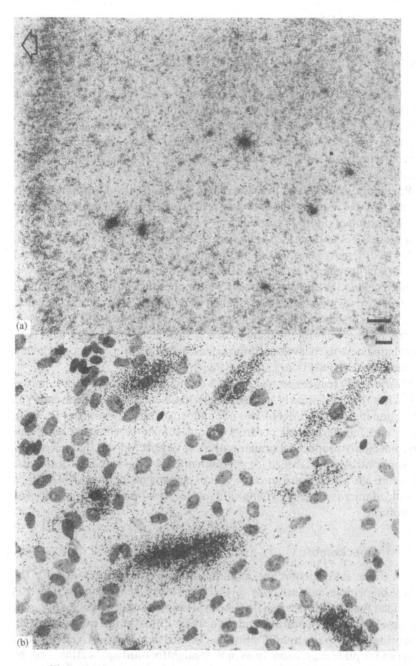

Figure 8. (a) Numerous neurones are heavily labelled following autoradiography in the rat cortex after *in situ* hybridisation with a NPY cRNA probe labelled with ^{32}P. Arrow points to the outer cortical layer. (b) *In situ* hybridisation of ANP mRNA in a culture of rat atrial myocytes using a ^{35}S-cRNA labelled probe. (a), (b) Counterstained with haematoxylin. Scale bars = 20 μm.

modulated by neural and endocrine controls. Prolactin synthesis and secretion by the pituitary gland is stimulated by oestrogen (MacLeod et al., 1969) at the level of gene transcription (Maurer et al., 1982). Thyroid status is known to influence levels of prolactin secretion (Peake et al., 1973) and presumably the expression of the gene. In this study we chose to examine prolactin gene expression in rats with varied oestrogen status or after thyroidectomy, in order to investigate whether changes take place at a cellular level.

Pregnant, lactating and ovariectomised female rats and normal controls were compared, and thyroidectomised and normal males were also included. Hybridisation of pituitary sections with a cRNA encoding prolactin mRNA was carried out as described elsewhere (Steel et al., 1988).

Large differences in intensity and frequency of labelled cells were seen in the various groups of rats. In normal female and male rats, prolactin cRNA-labelled cells in the anterior pituitary were numerous and evenly distributed, with individual cells intensely labelled (Fig. 9(a)). In pregnant females, however, prolactin mRNA was greatly reduced (Fig. 9(b)) due to very low oestrogen levels during pregnancy. In contrast, lactating females with high circulating oestrogen levels had very intense labelling over many cells, giving a much higher signal than controls (Fig. 9(c)).

Ovariectomy reduces oestrogen levels dramatically and in ovariectomised rats a few relatively weakly-labelled cells remained, although apparently the reduction in prolactin mRNA was not uniform in all lactotrophs (Fig. 9(d)). Male and female normal controls gave a similar labelling pattern, and thyroidectomy produced a similar pattern to ovariectomy in that fewer cells than in controls were labelled for prolactin mRNA (Fig. 9(e)). This technique therefore allows the distribution of cells expressing endocrine mRNAs to be examined microscopically, giving valuable information about levels of gene expression.

Conclusions

The introduction and application of *in situ* hybridisation for the study of peptide and mRNAs in the diffuse neuroendocrine system has provided further information on gene expression and the possible functions of these peptides in neural and endocrine cells. From the few examples given, it is evident that *in situ* hybridisation can be used in a variety of ways to study gene and peptide expression. In its most straightforward application it can be employed to localise mRNAs encoding known peptides and thus can be used as a useful adjunct to immunocytochemistry, and in many cases provides a more sensitive technique. It may also be used to map the distribution of newly discovered genes, or genes which appear in unexpected locations (e.g.

In situ hybridisation of peptide mRNAs   59

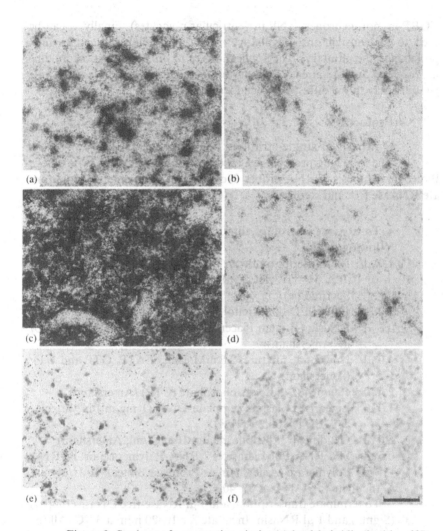

Figure 9. Sections of rat anterior pituitary gland hybridised with a ^{32}P-labelled prolactin probe for detection of mRNA. (a)–(d) female (e)–(f) male rats. (a) control (b) pregnant (c) lactating and (d) ovariectomised animals. (e) Thyroidectomised male rat and (f) tissue section from the same cryostat block as (e) pretreated with RNase prior to hybridisation. (a)–(f) Counterstained with haematoxylin. Scale bar = 20 μm.

CGRP in motoneurones, ANP in ventricular myocytes). Finally, it can be used to gain insight into regulatory events and cellular dynamics (e.g. prolactin in the pituitary). The applications of *in situ* hybridisation are thus far-reaching and are already leading to a better understanding of the rôle peptides play in the diffuse neuroendocrine system.

Protocols

Synthesis of single-stranded cRNA probes

The following protocol is a modification of that given by Promega Biotec for synthesis of RNA probes. Note that all equipment and solutions are RNAase free and sterilised.

- To a sterile microfuge tube, at room temperature, add in the following order:
- 4.0 μl 5 × transcription buffer
    (0.2 M Tris HCl, pH 7.5; 30 mM $MgCl_2$, 10 mM Spermidine)
- 2.0 μl 100 mM Dithiothreitol (DTT)
- 0.8 μl RNasin (Human Placental Ribonuclease Inhibitor): 25 U/μl (Amersham)
- 4.0 μl Nucleotide mixture: 2.5 mM each of ATP, GTP and UTP (Amersham)
- 2.4 μl 100 mM Cytidine triphosphate (CTP) (Amersham)
- 1.0 μl Linearised plasmid template DNA (1 mg/ml) in water or Tris-EDTA buffer (1 μg)
- 5.0 μl Cytidine ([^{32}P] triphosphate) 10 mCi/ml; Amersham)
- 0.5 μl SP6 RNA Polymerase, T7 RNA Polymerase or T3 RNA Polymerase (Amersham or Promega Biotec): 10 U/μl.
- Incubate for 1–1½ h at 37–40 °C.
- To terminate transcription, add 1 μl RNase-free DNase (1 μg/μl) (Sigma) and 1 μl RNasin. Incubate for 10–20 min at 37 °C. All the solutions mentioned in steps 1–3 are stored at −20 °C.
- Add 1 μl total RNA: 10 μg/μl (Sigma)
    175 μl diethyl pyrocarbonate (DEPC) water
    5 μl 4 M NaCl.

Extract with an equal volume (200 μl) of phenol/chloroform (1:1 v/v). Mix by vortexing. Separate the phases by centrifugation (5 min).

- Remove the upper aqueous phase (200 μl) and extract this with an equal volume (200 μl) of chloroform. Mix and spin as above.
- To the upper aqueous layer add 100 μl 7 M ammonium acetate

(2.5 M final concentration), 750 μl absolute ethanol ($-20°C$). Mix and leave at $-20°C$ overnight.
- Microfuge for 30 min. Discard the supernatant. Dry the RNA pellet under vacuum. When dry, dissolve the pellet in 20 μl of water for assessment of incorporation of radioactivity.
- Incorporation of radioactivity is estimated by determination of trichloroacetic acid (TCA)-precipitable counts.

1 μl labelled RNA probe
50 μl Bovine serum albumin (10 μg/μl)
1 μl 10%.
- Vacuum filter on GF/C (Glass microfibre paper, Whatman).
- Wash the filter twice with 10% TCA and twice with absolute ethanol. Dry the filter.
- Count the radioactivity on the filters and from this determine the % incorporation of radioactivity.

Preparation of small cRNA probes by limited alkaline hydrolysis

- After ethanol precipitation, spin down probe and dry under vacuum.
- Dissolve in 80 μl sterile RNase-free water.
- Add 10 μl of 400 mM $NaHCO_3$ (final concentration = 40 mM) and 10 μl of 600 mM Na_2CO_3 (final concentration = 60 mM).
- Heat at 60°C for required hydrolysis time.

Hydrolysis time is calculated from the following formula:
$$t = \frac{L_o - L_f}{K(L_o L_f)}$$
where t = time for hydrolysis in minutes
L_o = original probe length in kilobases
L_f = final probe length in kilobases
K = (rate constant) = 0.11

- Stop hydrolysis by adding
10 μl sodium acetate (0.1 M final concentration)
5 μl acetic acid (0.5% v/v final concentration)
(final pH should be 6.0).
- Add 2.5 volumes ethanol (290 μl) and leave at $-20°C$ to precipitate the RNA.

Fixation

Where applicable tissues should be fixed by perfusion of fixative transcardially. When this is not possible, i.e. in cases of surgical samples, care should be taken to ensure the tissues are sliced thinly (0.5 cm thick) and are immersed in an adequate volume (usually 25 × volume of tissue) of fixative.

Preparation of fixative – 4% paraformaldehyde (0.1 M, pH 9.5)
For 1 litre

- Dissolve 38.2 sodium borate and 4.0 g sodium hydroxide in 1 litre distilled water and heat to 60–65 °C.
- Add 40 g paraformaldehyde, dissolve and cool to room temperature.
- Adjust pH to 9.5 with 1.0 M sodium hydroxide and cool to 4 °C.
- This fixative should be freshly prepared and not allowed to stand for more than 8 h. Alternatively, frozen aliquots can be successfully used.

Prehybridisation treatment
- Adhere sections/cultures to prepared slides and bake overnight at 43 °C.
- Rehydrate in phosphate-buffered saline (PBS) (0.1 M, pH 7.2) for 5–10 min.
- Immerse in glycine/PBS (0.1 M) for 5 min.
- Soak slides in 0.3% Triton X-100 in PBS for 10–15 min to permeabilise.
- Wash thoroughly with PBS 3 × 5min.
- Incubate with Proteinase K (1 μg/ml) in Tris HCl (0.1 M, pH 8.0) and EDTA (50 mM, pH 8) for 20 min at 37 °C to deproteinise.
- Stop reaction by immersion in freshly made 4% paraformaldehyde in PBS (0.1 M, pH 7.2).
- Acetylate with acetic anhydride (0.25% v/v) in triethanolamine (0.1 M, pH 8.0) for 10 min.
- Prehybridise in formamide (50% v/v) in 2 × standard sodium citrate at 37 °C for 15–45 min.

Hybridisation
- Drain excess formamide solution from slides and apply 20 μl of hybridisation mixture at 42 °C containing radiolabelled probe ($5 \times 10^3 - 5 \times 10^6$ cpm per section/culture well) diluted in hybridisation buffer.
- Cover the sections/cultures with dimethyl dichloro silane-coated cover-slips and incubate in a humid atmosphere at 42 °C for 12–18 h.

Post-hybridisation washing
- Immerse slides in 4 × standard sodium citrate (SCC) to remove coverslips.

- Wash slides with gentle agitation in 4 × SCC at 42°C, 3 × 20 min.
- Immerse in RNAase A (20 μg/ml) in NaCl (0.5 M) Tris-HCl (pH 8.0) EDTA (1 mM, pH 8.0) at 42°C for 30 min.
- Wash sections in descending concentrations of SSC; 2 × SSC, 0.1 × SSC and 0.05 SSC 30 min each at 42°C.
- Dehydrate with 70%, 90% and two changes of 100% ethanol containing 0.3 M ammonium acetate 10 min each at room temperature. Air dry.

Autoradiography

(a) Film method:
- Expose labelled sections to radiation sensitive film (e.g. Amersham); ^{32}P labelled sections for 24 h, ^{35}S labelled sections for 48 h, in darkness at 4°C. Labelled sections must be exposed with radioactive standard, either ^{32}P or ^{35}S.
- Develop X-ray film in Kodak D 19 developer prepared and used according to manufacturer's instructions (5 min at 18°C).
- Rinse in distilled water containing 0.1% glacial acetic acid (30 s).
- Fix with AMFIX (photographic fixer) solution for 5 min at 18°C.
- Wash under running tap water for 15 min.

(b) Dipping method:
- Dip sections in an autoradiographic emulsion (Kodak NTB-2 or Ilford K-5) diluted 1:1 with double distilled water at 42°C.
- Air dry for 3 h.
- Expose slides in a light-proof box for 2–5 days depending on the radioisotope used (2–3 days for ^{32}P, 5 days for ^{35}S).
- Develop with Kodak D 19 for 3 min at 18°C.
- Rinse in distilled water for 30 s.
- Fix with AMFIX solution for 4–5 min, 18°C.
- Wash section under running tap water for 10–15 min.
- Dehydrate, clear and mount with DPX.

References

Allen, J. M. & Bloom, S. R. (1986). Review: Neuropeptide Y: A putative neurotransmitter. *Neurochemical International*, **8**, 1–8.

Amara, S. G., Arriza, J. L., Leff, S. E., Swanson, L. W., Evans, R. M. & Rosenfeld, M. G. (1985). Expression of brain messenger RNA encoding a novel neuropeptide homologous to calcitonin gene-related peptide. *Science*, **229**, 1094–7.

Amara, S. G., Jonas, V., Rosenfeld, M. G., Ong, E. S. & Evans, R. M. (1982). Alternative RNA-processing in calcitonin gene expression generates mRNAs encoding different polypeptide products. *Nature (London)*, **298**, 240–4.

Atlas, S. A., Klienert, H. D., Camargo, M. J., Janaszewicz, A., Sealey, J. E., Laragh, J. H., Schilling, J. W., Lewicki, J. A., Johnson, L. K. & Maack, T. (1984). Purification, sequencing and synthesis of natriuretic and vasoactive rat atrial peptide. *Nature (London)*, **309**, 717–19.

Chan-Palay, V., Allen, Y. S., Lang, W., Haesler, U. & Polak, J. M. (1985a). I. Cytology and distribution in human cerebral cortex of neurons immunoreactive with antisera against neuropeptide Y. *Journal of Comparative Neurology*, **238**, 382–9.

Chan-Palay, V., Lang, W., Allen, Y. S., Haesler, U., & Polak, J. M. (1985b). II Cortical neurons immunoreactive with antisera against neuropeptide Y are altered in Alzheimer's type dementia. *Journal of Comparative Neurology*, **238**, 390–400.

Chapeau, C., Gutowska, J., Schiller, P. W., Milne, R. W., Thibault, G., Garcia, R., Genest, J. & Cantin, M. (1985). Localization of immunoreactive synthetic atrial natriuretic factor (ANF) in the heart of various animal species. *Journal of Histochemistry & Cytochemistry*, **33**, 541–50.

Cox, K. H., DeLeon, D. V. & Angerer, L. M. (1984). Detection of mRNA in sea urchin embryos by *in situ* hybridisation using a symmetric RNA probe. *Developmental Biology*, **101**, 485–502.

Denny, P., Hamid, Q., Krause, J. E., Polak, J. M. & Legon, S. (1988). Oligoriboprobes: Tools for *in situ* hybridisation. *Histochemistry*, **89**, 481–3.

Du Vigneaud (1956). Hormones of the posterior pituitary gland: oxytocin and vasopressin. *Harvey Lectures, 1954–5*. New York: Academic Press.

Facer, P., Hamid, Q. A., Bishop, A. E., Denny, P., Legon, S., Harmar, A. J., Goodman, R. H. & Polak, J. M. (1988). Neuropeptide m-RNAs in the gastrointestinal nervous system of man and rat localised by *in situ* hybridisation. *Regulatory Peptides*, **22**, 403.

Feyrter, F. (1938). *Uber diffuse endokrine epiteliale Organe*. Liepzig: J. A. Barth.

Fontaine, B., Klarsfield, A., Hökfelt, T. & Changeux, J.-P. (1986). Calcitonin gene-related peptide, a peptide present in spinal cord motoneurones, increases the number of acetylcholine receptors in primary cultures of chick embryo myotubes. *Neuroscience Letters*, **71**, 59–65.

Gall, J. C. & Pardue, M. L. (1969). Formation and detection of RNA–DNA hybrid molecules in cytological preparations. *Proceedings of the National Academy of Sciences, USA*, **63**, 378–83.

Gardener, D. G., DeSchepper, C. F., Ganong, W. F., Hane, S., Fiddes, J., Baxter, J. D. & Lewicki, J. D. (1986). Extra-atrial expression of the gene for atrial natriuretic factor. *Proceedings of the National Academy of Sciences, USA*, **83**, 6697–701.

Gibson, S. J., Polak, J. M., Bloom, S. R., Sabate, I. M., Mulderry, P. M., Ghatei, M. A., McGregor, G. P., Morrison, J. F. B., Kelly, J. S., Evans, R. M. & Rosenfeld, M. G. (1984). Calcitonin gene-related peptide

(CGRP) in the spinal cord of man and of eight other species. *Journal of Neuroscience*, **4**, 3101–11.
Gibson, S. J., Polak, J. M., Giaid, A., Hamid, Q. A., Kar, S., Jones, P. M., Denny, P., Legon, S., Amara, S. G., Craig, R. K., Bloom, S. R., Penketh, R. J. A., Rodek, C., Ibrahim, N. B. N. & Dawson, A. (1988). Calcitonin gene-related peptide messenger RNA is expressed in sensory neurones of the dorsal root ganglia and also in spinal motoneurones in man and rat. *Neuroscience Letters*, **91**, 283–8.
Hamid, Q., Adam, C., Terenghi, G. & Polak, J. M. (1988). The potential of biotinylated cRNA probes for the investigation of gene expression in the diffuse neuroendocrine system using *in situ* hybridisation. *Journal of Pathology*, **154**, 84A.
Hamid, Q. A., Wharton, J., Terenghi, G., Hassall, C. J. S., Aimi, J., Taylor, K. M., Nakazato, H., Dixon, J. E., Burnstock, G. & Polak, J. M. (1987). Localization of atrial natriuretic peptide mRNA and immunoreactivity in the rat heart and human atrial appendage. *Proceedings of the National Academy of Sciences, USA*, **84**, 7315–18.
Henry, J. L. (1982). Relation of substance P to pain transmission: neurophysiological evidence. In *Substance P in the Nervous System*, (eds Porter, R. & O'Connor, M.), pp. 206–24, London: Pitman.
Henry, J. L. (1977). Substance P and pain: a possible relation in afferent transmission. In *Substance P* (eds von Euler, U. S. & Pernow, B.), pp. 232–40, New York: Raven Press.
Hökfelt, T., Kellerth, J.-O., Nilsson, G. & Pernow, G. (1975). Experimental immunohistochemical studies on the localization and distribution of substance P in cat primary sensory neurons. *Brain Research*, **100**, 235–52.
Huang, W. M., Gibson, S. J., Facer, P., Gu, J. & Polak, J. M. (1983). Improved section adhesion for immunocytochemistry using high molecular weight polymers of L-Lysine as a slide coating. *Histochemistry*, **77**, 275–9.
Jordan, C. C. & Oehme, P. (1985). *Substance P. Metabolism and Biological Actions*. London & Philadelphia: Taylor & Francis.
Kanagawa, K., Tawaragi, T., Oikawa, S., Mizuno, A., Sakuragawa, Y., Nakazato, H., Fukuda, A., Minamino, N. & Matsuo, H. (1984). Identification of rat atrial natriuretic polypeptide and characterization of the cDNA encoding its precursor. *Nature (London)*, **312**, 152–5.
Kikuchi, K., Nakao, K., Hayashi, N., Morii, N., Sugawara, A., Sakamoto, M., Imura, H. & Mikawa, H. (1987). Ontogeny of atrial natriuretic polypeptide in the human heart. *Endocrinology*, **115**, 211–17.
Larsson, L.-I., Fahrenkrug, J., Schaffalitzky, de Muckadell, O., Sundler, F., Hakanson, R. & Rehfeld, J. F. (1976). Localization of vasoactive intestinal polypeptide (VIP) to central and peripheral neurons. *Proceedings of the National Academy of Sciences, USA*, **73**, 3197–200.
Lattion, A. L., Michel, J. B., Arnauld, E., Corrol, P. & Soubrier, F. (1986). Myocardial recruitment during ANF mRNA increase with volume overload in the rat. *American Journal of Physiology*, **251**, H890–6.

MacLeod, R. M., Abad, A. & Eidson, L. L. (1969). *In vivo* effects of sex hormones on the *in vitro* synthesis of prolactin and growth hormone in normal and pituitary tumour-bearing rats. *Endocrinology*, **84**, 1475–83.

Maki, M., Takayanagi, R., Misono, K. S., Pandy, K. N., Tibbettes, C. K. & Inagami, T. (1984). Structure of rat atrial natriuretic factor precursor deduced from cDNA sequence. *Nature (London)*, **309**, 722–4.

Matteoli, M., Haimann, C., Torri-Tarelli, F., Polak, J. M., Cecceralli, B. & De Camilli, P. (1988). Differential effect of α-latrotoxin on exocytosis from small synaptic vesicles and from CGRP-containing large dense core vesicles at the frog neuromuscular junction. *Proceedings of the National Academy of Sciences, USA*, **85**, 7366–70.

Maurer, R. A. (1982). Estradiol regulates the transcription of the prolactin gene. *Journal of Biological Chemistry*, **257**, 2133–6.

Morris, H. R., Panico, M., Etienne, T., Tippins, J., Girgis, S. I. & MacIntyre, I. (1984). Isolation and characterization of human calcitonin gene-related peptide. *Nature (London)*, **308**, 746–8.

Mulderry, P. K., Ghatei, M. A., Spokes, R. A., Jones, P. M., Pierson, A. M., Hamid, Q. A., Kanse, S., Amara, S. G., Burrin, J. M., Legon, S., Polak, J. M. & Bloom, S. R. (1988). Differential expression of α-CGRP and β-CGRP by primary sensory neurones and enteric autonomic neurons. *Neuroscience*, **25**, 195–205.

Nawa, H., Hirose, T., Takashima, H., Inyami, S. & Nakanishi, S. (1983). Nucleotide sequences of cloned cDNAs for two types of bovine brain substance P precursors. *Nature (London)*, **306**, 32–6.

Nawa, H., Kotani, H. & Nakanishi, S. (1984). Tissue-specific generation of the preprotachykinin mRNAs from one gene by alternative RNA splicing. *Nature (London)*, **312**, 729–34.

New, H. V. & Mudge, A. W. (1986). Calcitonin gene-related peptide regulates muscle acetylcholine receptor synthesis. *Nature (London)*, **323**, 809–11.

Nilsson, G., Larsson, L.-I., Brodin, E., Pernow, P. & Sundler, F. (1975). Localisation of substance P-like immunoreactivity in mouse gut. *Histochemistry*, **43**, 97–9.

Oikawa, S., Imai, M., Ueno, A., Tanaka, S., Noguchi, T., Nakazato, H., Kanagawa, K., Fukuda, A. & Matsuo, M. (1984). Cloning and sequence analysis of cDNA encoding a precursor for human atrial natriuretic polypeptide. *Nature (London)*, **309**, 724–6.

Pardue, M. L. (1985). *In situ* hybridisation. In *Nucleic Acid Hybridisation* (eds Hames, B. D. & Higgin, S. J.), pp. 179–202. Oxford, Washington: IRL Press.

Peake, G. T., Birge, C. A. & Daughaday, W. H. (1973). Alterations of radioimmunoassayable growth hormone and prolactin during hypothyroidism. *Endocrinology*, **92**, 487–93.

Pearse, A. G. E. (1969). The cytochemistry and ultrastructure of polypeptide-hormone producing cells of the APUD series, and the embryonic, physiologic and pathologic implications of the concept. *Journal of*

Histochemistry & Cytochemistry, **17**, 303–13.
Pearse, A. G. E., & Polak, J. M. (1975). Immunocytochemical localisation of substance P in mammalian intestine. *Histochemistry,* **41**, 373–5.
Polak, J. M. & Bloom, S. R. (1982). Distribution and tissue localization of VIP in the central nervous system and in seven peripheral organs. In *Vasoactive Intestinal Peptide* (ed. Said, S. I.), pp. 107–20. New York: Raven Press.
Rosenfeld, M. G., Mermod, J. J., Amara, S. G., Swanson, L. W., Sawchenko, P. E., Rivier, J., Vale, W. W. & Evans, R. M. (1983). Production of a novel neuropeptide encoded by the calcitonin gene via tissue-specific RNA processing. *Nature (London),* **304**, 129–35.
Said, S. I. (1984). Vasoactive intestinal polypeptide (VIP) current status. *Peptides,* **5**, 143–50.
Said, S. I. & Mutt, V. (1970). Polypeptide with broad biological activity: isolation from small intestine. *Science,* **169**, 1217–18.
Salt, T. E. & Hill, R. G. (1983). Neurotransmitter candidates of somatosensory primary afferent fibres. *Neuroscience,* **10**, 1083–103.
Singer, R. H., Lawrence, J. B. & Villnave, C. (1986). Optimization of *in situ* hybridisation using isotopic and non-isotopic detection methods. *Biotechniques,* **4**, 230–50.
Steel, J. H., Hamid, Q., Van Noorden, S., Jones, P., Denny, P., Burrin, P., Legon, S., Bloom, S. R. & Polak, J. M. (1988). Combined use of *in situ* hybridisation and immunocytochemistry for the investigation of prolactin gene expression in immature, pubertal, pregnant, lactating and ovariectomised rats. *Histochemistry,* **89**, 75–80.
Stevens, J. G., Wagner, E. K., Devi-Rao, G. B., Cook, M. L. & Feldman, L. T. (1987). RNA complementary to a herpes virus gene mRNA is prominant in latently infected neurons. *Science,* **235**, 1056–9.
Takami, K., Kawai, Y., Shiosaka, S., Lee, Y., Girgis, S., Hillyard, C. J., MacIntyre, I., Emson, P. C. & Tohyama, M. (1985). Immunohistochemical evidence for the coexistence of calcitonin gene-related peptide- and choline acetyltransferase-like immunoreactivity in neurons of the rat hypoglossal, facial, and ambiguus nuclei. *Brain Research,* **328**, 386–9.
Tatemoto, K. (1982). Neuropeptide Y: complete amino acid sequence of the brain peptide. *Proceedings of the National Academy of Sciences, USA,* **79**, 5485–9.
Terenghi, G., Polak, J. M., Hamid, Q., O'Brian, E., Denny, P., Legon, S., Dixon, J., Minth, C. D., Palay, S. L., Yasargil, G. & Chan-Palay, V. (1987). Localization of neuropeptide Y mRNA in neurons of human cerebral cortex by means of *in situ* hybridisation with a complementary RNA probe. *Proceedings of the National Academy of Sciences, USA,* **84**, 7315–18.
Yamanaka, M., Greenberg, B., Johnson, L., Seilhamer, J., Brewer, M., Friedmann, T., Miller, J., Atlas, S., Laragh, J., Lewicki, J. & Fiddes, J. (1984). Cloning and sequence analysis of the cDNA for the rat atrial natriuretic factor precursor. *Nature (London),* **309**, 719.

Robert C. Angerer, Susan D. Reynolds, Julia Grimwade, David L. Hurley, Qing Yang, Paul D. Kingsley, Michael L. Gagnon, James Palis and Lynne M. Angerer

Contributions of the spatial analysis of gene expression to the study of sea urchin development

Early motivations

In 1977 we (L. M. A. and R. C. A.) submitted our first research proposal to the National Institutes of Health. The most speculative part was to use *in situ* hybridisation techniques to determine the distribution of individual mRNAs in sea urchin embryos. We suggested that information on spatial distribution would be useful in approaching three questions. First, we were puzzled that molecular assays of gene expression in whole embryos had led to the generalisation that most mRNAs are found throughout development (reviewed by Davidson, 1976, 1986) and that a relatively small percentage show large changes in abundance on a per embryo basis. The missing information required to interpret these observations was the extent to which genes are spatially regulated in different tissues or lineages. Second, we hoped to identify sets of mRNAs whose expression is co-ordinated in time and space, so that the mechanism of that co-ordination could be investigated. Third, we suggested that individual spatially restricted mRNAs could serve as molecular markers to construct a fate map of the embryo, to determine when cells commit to specific patterns of gene expression, and possibly to identify maternal RNAs localised in the egg. We proposed a long-term commitment to 'localise the time and site of synthesis of individual mRNA species in developing embryos'. A dozen years seems an appropriately long time, and we take this opportunity to review the contribution of information gained from *in situ* hybridisation studies to the questions we initially anticipated, as well as to some we did not.

Methodology

We first used polyU probes to compare several different fixation and prehybridisation treatments (Angerer & Angerer, 1981). This work clearly showed that the cross-linking fixative, glutaraldehyde, in combination with mild protease digestion as suggested by Brahic & Haase (1978), provided better retention of target RNA and higher signals than could be achieved with precipitating fixatives such as ethanol–acetic acid.

Our first detection of individual mRNAs (Venezky et al., 1981) used a cloned sequence containing one copy of the 6.8 kb tandem repeat encoding early histone mRNAs (Overton & Weinberg, 1978). We soon found that signals provided per pg of target sequence were much lower for DNA probes containing both sense and antisense strands than were those for the 'antisense' polyU probe. The simplest explanation was that reassociation of self-complementary probes limited their penetration to target RNAs in the tissue. After several months of frustration from trying to strand-separate DNA probes, a breakfast conversation with Dr Tom Maniatis revealed a new vector system for cloning and *in vitro* transcription of antisense RNA probes using Sp6 promoter and polymerase developed by Butler & Chamberlin (1982). Within several months we had made a direct comparison of single- and double-stranded RNA probes, which showed that antisense RNAs provided about 8-fold higher signals. Kathleen Cox, who was then a graduate student, made a detailed study to determine optimal values for major variables such as probe concentration, hybridisation temperature and time (Cox et al., 1984). An important aspect of this early work was the comparison of signals provided by *in situ* hybridisation to those obtained with standard solution hybridisation methods. This demonstrated that quantitatively accurate estimates of relative signals over areas of different mRNA target density can provide measures of relative message concentration, even if saturation of targets is not achieved (Angerer & Angerer, 1981; Cox et al., 1984).

We continue to use almost exclusively the combination of tissue fixed with glutaraldehyde, limited pre-hybridisation digestion with proteinase K, and radioactively labelled antisense RNA probes, following essentially our original protocol (Cox et al., 1984). As we suggested then, the major difficulty in establishing conditions for *in situ* hybridisation to different tissues has been in identifying optimal methods for tissue preparation. An unexpected hazard was the fact that some individual steps reported to be beneficial in protocols worked out by various laboratories have proved to be disastrous in combination. We have discussed previously the factors to be considered in developing *in situ* methods for new systems (most completely in Angerer et al., 1987b; see also Angerer et al., 1987a and Angerer et al., 1987c). We favour RNA probes because they provide reliably high signals and because, unlike symmetric probes, they form 1:1 duplexes with RNA targets that assure reproducible quantitation. However, nick-translated DNA probes and formaldehye-fixed tissue have been employed effectively by others (e.g. as first documented for *Drosophila* embryos, Hafen et al., 1983), and these alternatives have been discussed most extensively by Lawrence & Singer (1985; see also Singer *et al.*, 1987).

Commitments to specific patterns of gene expression in the early sea urchin blastula

Over the next few years we used *in situ* hybridisation to determine the spatial distributions of a variety of mRNAs isolated in recombinant clones by colleagues in other laboratories. The first studies described an unusual accumulation and restriction of immense quantities of histone mRNA in egg pronuclei, a phenomenon whose molecular mechanism is yet to be elucidated (Venezky *et al.*, 1981; Showman *et al.*, 1982; DeLeon *et al.*, 1983). Comparison of the distributions of early and late histone variant mRNAs answered a long-standing question about the potential developmental significance of this switch: both early and late variant mRNAs are contained in the same cells during the switch, without any indication of cell-lineage or tissue specificity (Angerer *et al.*, 1985). Furthermore, as discussed below, early blastomeres commit to lineage- and tissue-specific patterns of mRNA accumulation long before late variant proteins comprise a significant fraction of total histone synthesis. Thus, the switch appears to serve primarily as a mechanism for quantitative regulation of histone mRNA synthesis, with the many early genes producing high levels of histone message during cleavage and then being shut down in the early blastula, while the few late genes continue to produce cell-cycle regulated mRNAs at much lower rates in dividing cells through pluteus stage (Cox *et al.*, 1984).

The first non-histone mRNA we examined was Spec1 (*Strongylocentrotus purpuratus* ectoderm), identified by Dr William Klein's laboratory as an ectoderm-enriched mRNA of unknown function (Bruskin *et al.*, 1982). This mRNA was found to be restricted to the aboral ectoderm of the pluteus larva (Lynn *et al.*, 1983). The distribution of this first tissue-specific mRNA substantiated our proposal to use mRNAs to improve the fate map of the early embryo because it delineated the presumptive oral and aboral regions on the histologically undifferentiated ectoderm of the blastula and early gastrula. This division was missing from the fate map constructed from classic dye marking experiments (for example, see Hörstadius, 1983, p. 16; for recent additions to the fate map of ectoderm, see Cameron *et al.*, 1987).

The Spec1 pattern also raised new questions. As shown in Figure 1, the fact that Spec1 mRNA is highly enriched in the presumptive aboral ectoderm of the blastula showed that cells of the future oral and aboral ectoderm regions commit to different patterns of gene expression long before the morphological consequences of such commitments are revealed at late gastrula stage.

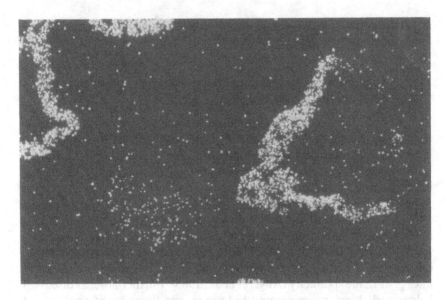

Figure 1. mRNA distributions foreshadow differentiation. *In situ* hybridisation with a probe for Spec1 mRNA shows that this message accumulates only in aboral ectoderm of the pluteus larva (at right). Accumulation of Spec1 mRNA is detectable 2 days earlier as shown by the grazing section of a hatching blastula shown at left.

We wondered whether similar early commitment also occurred in other tissues or lineages, when region-specific genes might first be activated, and whether this happened at distinct times for different sets of cells. Analysis of *in situ* hybridisation patterns of relatively few additional mRNAs showed that six major territories of specific mRNA accumulation are defined by mesenchyme blastula stage, as listed in Table 1 (reviewed by Angerer & Davidson, 1984 and Davidson, 1989). These territories, and their derivatives in embryos at later stages, are illustrated in Figure 2 and include oral and aboral ectoderm, primary and secondary mesenchyme, endoderm and the small micromeres. Primary mesenchyme and aboral ectoderm territories give rise to single histological types of terminally differentiated cells in the unfed pluteus larva. The other territories include precursors to several different cell types. For example, in the pluteus larva the distributions of different cytoplasmic actin mRNAs (see Cox *et al.*, 1986) and those of various antigens (McClay & Wessel, 1985) distinguish between foregut and mid+hindgut and muscle actin mRNA is restricted to a few secondary mesenchyme cells that form muscle fibres around the esophagus (Cox *et al.*, 1986). Some territories are identified by unique accumulation(s) of one or more mRNAs, as is discussed for aboral ectoderm cells on page 80. Other

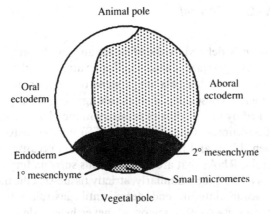
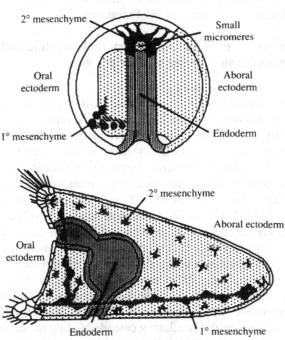

Figure 2. Major territories of the early blastula, and their derivatives at later stages of development. (a) Fate map of blastula stages before ingression of primary mesenchyme cells (about 10 to 20 h). The map is similar at mesenchyme blastula stage, except primary mesenchyme cells have ingressed into the blastocoel. (b) Gastrula stage embryo (about 36–48 h). (c) Pluteus larvae (about 72 h). Secondary mesenchyme cells produce pigment cells dispersed throughout the ectoderm of the pluteus, and a few muscle cells around the esophagus (not shown). Derivatives of the small micromeres contribute to the coelomic rudiments on either side of the esophagus (not shown).

territories are currently defined by the presence and/or absence of a specific combination of messages (see Table 1). For example, eight cells at the extreme vegetal pole of the blastula are progeny of the four small daughter cells of the micromeres. These cells do not divide again until late pluteus stage and can be identified by immunohistochemistry in the blastula because they retain high concentrations of cleavage stage (CS) variant histone proteins (Pehrson & Cohen, 1986) and by *in situ* hybridisation because they do not accumulate several mRNAs that are found in cells surrounding them in the vegetal plate (Cox *et al.*, 1986). Similarly, strictly tissue-specific messages do not appear to accumulate in endoderm until gastrulation is almost completed; at earlier stages this region is, nevertheless, definable by the presence and absence of a combination of messages.

Tissue-specific messages in the very early blastula and early synthesis of tissue-specific nuclear proteins

In pursuit of regulatory genes

As for other embryos, a major goal in the study of sea urchin development is to understand the molecular mechanisms that set up spatially restricted patterns of gene expression in the very early embryo. Because sea urchins are not amenable to standard genetic analysis, one route being pursued is to work toward early events starting from regulatory molecules that operate on tissue-specific genes. *In situ* hybridisation has contributed significantly to this effort, first by identifying appropriate tissue-specific genes (Lynn *et al.*, 1983; Cox *et al.*, 1986; Benson *et al.*, 1987) and subsequently as the assay for appropriate tissue-specific expression of 'wild type' and deleted constructs in embryos developing from eggs microinjected with recombinant DNA test genes (e.g. the aboral ectoderm-specific genes, actin CyIIIa (Hough-Evans *et al.*, 1987; Franks *et al.*, 1988*a,b*; Calzone *et al.*, 1988) and Spec1 (W. H. Klein and coworkers, personal communication); and the primary mesenchyme-specific Sm50 gene (Sucov *et al.*, 1988)).

The number of steps in the regulatory cascade leading to tissue-specific gene expression is unknown. Immediate regulation of aboral ectoderm-specific genes such as actin CyIIIa and Spec1, which encode abundant proteins of differentiated cells, may be several steps removed from information originally localised in the egg. If a round(s) of gene expression were prerequisite to activation of such 'late' tissue-specific genes, then it would have to occur before these late mRNAs begin to accumulate. To search for such early mRNAs, several years ago we began screening for the earliest tissue-specific mRNAs expressed in the sea urchin embryo (Reynolds *et al.*, in preparation). As an independent alternative route to potential regulatory molecules, we also began an analysis of the metabolism of tissue-specific nuclear proteins (Grimwade *et al.*, in preparation).

Table 1. *mRNAs expressed in major territories of the mesenchyme blastula stage sea urchin embryo*

	Aboral ectoderm	Oral ectoderm	Endoderm	2° mesenchyme	1° mesenchyme	Sm. micromeres
actin CyIIIa	++	–	–	–	–	–
Spec1,2a/c	++	–	–	–	–	–
SpARS	++	–	–	–	–	–
SpHbox1	++	–	–	–	–	–
SpEGF2	++	+	–	–	–	–
Spec3	++	++	–	–	–	–
actin CyI,IIb	–	++	++	++	+/–	–
actin CyIIa	–	–	–	++	+	–
collagen	–	–	–	+	++	–
Sm50	–	–	–	–	+	–

All mRNAs are discussed, and references provided in the text.

Tissue-specific mRNAs transiently expressed at very early blastula (VEB)

To identify recombinant DNA clones representing spatially restricted mRNAs expressed in the very early blastula, we prepared a cDNA library from pooled polysomal RNAs of 9, 12 and 15-h embryos. The first screening criterion imposed was temporal: we used substracted cDNA probes to discard clones also represented in RNA from pluteus larvae, thus eliminating messages whose accumulation simply accompanies differentiation and only reflects earlier commitments. Standard developmental RNA blot analysis was then used to sort these early mRNA clones into two sets. The first set contains sequences also represented in maternal RNA, which have not been pursued because they are likely to be weighted heavily toward genes involved in rapid cell replication of cleavage (Grainger et al., 1986). The second set of clones represents mRNAs that are transiently expressed in very early blastula, which we call the VEB mRNA set (Reynolds et al., in preparation).

Messages encoded by the five VEB genes identified share interesting features of expression. First, their temporal regulation is distinctive. They begin to accumulate around 9–12 h, rapidly reach peak abundance by 12–15 h, and then decay equally rapidly until at 18–20 h (hatching blastula) they are present at only a small fraction of their maximum levels. Peak abundance of different VEB messages varies over a range of at least 50-fold. The rapid decay requires a half-life much shorter than that of average mRNAs measured during the blastula–gastrula period (5.7 h; Galau et al., 1977). Furthermore, because the screening protocol selected only against mRNAs expressed at pluteus stage (72 h), it is remarkable that all these mRNAs should be down-regulated so rapidly. This temporal co-ordination suggests that the VEB genes may have related functions. Second, *in situ* hybridisation of eight of the VEB mRNAs shows that *all* are spatially restricted. Although we have not completed a detailed study, it is clear that VEB mRNAs are absent from vegetal pole cells as shown by comparison of VEB signals to those for Spec3 mRNA in adjacent sections, as is illustrated in Figure 3(a) and (b). With all VEB probes examined to date, the labelled area in 12 and 15 h embryos (150–200 cells) is always sufficiently large to include all presumptive ectoderm, which encompasses about 70% of the embryo volume. It is likely that at least some of the VEB mRNAs are also present at these early times in precursors to endoderm and secondary mesenchyme, because many sections show only small unlabelled areas (Fig. 3(c)). Since each of the probes for individual VEB mRNAs labels only one contiguous region at 12–15 h, none of these mRNAs is expressed in two lineages separated by cells that do not accumulate that message (e.g. presumptive

Spatially regulated sea urchin genes

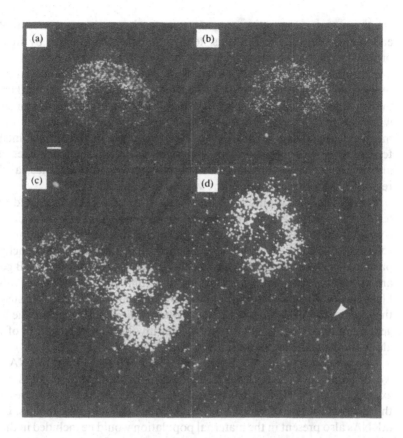

Figure 3. Spatially restricted mRNAs transiently expressed at very early blastula. (a) Hybridisation of a probe for Spec3 mRNA to a section of a 15-h embryo, at which stage presumptive ectoderm is labelled, but cells of vegetal lineages are not. (b) Immediately adjacent section hybridised with a probe for one of the VEB mRNAs, demonstrating that the unlabelled region lies at the vegetal pole. (c) Hybridisation of the same VEB probe as in (a) and (b) to section of two 12-h embryos, illustrating that the unlabelled region appears to be smaller at earlier stages. (d) As in (c), but hybridisation was with a different VEB probe. The arrowhead points to an unlabelled 20-h blastula. Hybridisation was carried out according to references cited in the text, using saturating concentrations of probe labelled with ^{35}S to a specific activity of 3×10^8 dpm/μg. Exposure was 6 days.

ectoderm and presumptive secondary mesenchyme, but not presumptive endoderm). The only region consistently unlabelled at all stages by all VEB probes corresponds to descendents of the 16-cell stage micromeres (skeletogenic primary mesenchyme and small micromeres). Detailed spatial maps using combinations of serial and adjacent section analyses with markers of known tissue specificity at blastula stage will be required to demonstrate what we now suspect – that the VEB messages are a nested set whose members are always expressed in ectoderm and incrementally include territories of endoderm and presumptive secondary mesenchyme. It is already clear, however, that as the messages decrease in abundance, labelled regions retract first from both the vegetal and animal poles.

The VEB clones represent the *first* region-specific genes activated after fertilisation. Lineages in the vegetal hemisphere that are restricted to ectoderm or endoderm + secondary mesenchyme do not separate until 64-cell stage (about 8–9 h), while endoderm and secondary mesnenchyme presumably separate at least one division later (about 11 h). The VEB genes are therefore activated very close to the time that cleavage partitions blastomeres with different fates along the animal–vegetal axis. We suspect that among the VEB proteins, or the proteins that *directly* regulate the VEB genes, are important regulatory molecules involved in specification of fate along the animal–vegetal axis.

Two observations about patterns *not* observed for the VEB mRNAs are worth comment. First, the fact that only micromeres derivatives never express any of the VEB mRNAs examined to date probably correlates with the fact that the fate of micromeres is determined by maternal factors. Early mRNAs also present in the maternal population would be included in the set of clones we have not examined. Second, at times of peak accumulation of VEB mRNAs, we have not detected oral–aboral differences in the presumptive ectoderm, although Cameron et al., (1987) have shown that blastomeres restricted to oral and aboral fates separate in the animal hemisphere of the embryo at 32-cell stage (discussed in Davidson, 1989). It is therefore possible that analysis of the VEB genes and of the regulation of aboral ectoderm-specific messages will lead in different directions, and help elucidate mechanisms of regulation of the animal–vegetal, and oral–aboral axes, respectively.

Tissue-specific nuclear proteins

An alternative route to candidates for regulatory molecules is through identification of tissue-specific nuclear proteins. Servetnick & Wilt (1987) have shown that the greatest number of changes in synthesis of nuclear proteins of whole embryos occur at the 64–200-cell stage. We have analysed nuclear proteins synthesised at various stages of embryogenesis using labelling protocols designed to determine times of synthesis, nuclear accumulation, and stability; tissue fractionation methods (McClay, 1986) have been used to separate plutei into fractions enriched for ectoderm or endoderm + mesenchyme (Grimwade *et al.*, in preparation). Of 120 proteins examined, 9 are ectoderm-specific, 8 are endoderm + mesenchyme-specific, and 18 show different tissue specificity at pluteus depending on the time of their synthesis. Because this analysis pools a number of different cell types, it undoubtedly underestimates the fraction of proteins that is cell type-specific.

Most important, in context of this discussion, are three observations about these tissue-specific nuclear proteins. First, most proteins highly enriched in one or the other tissue fraction of the pluteus begin to be synthesised at much earlier stages, usually as early as 16-cell stage; they are likely, therefore, to be encoded at least in part by maternal mRNAs. Second, some of these proteins are very stable; many of them can be shown after pulse labelling at 16-cell stage to persist for three days to pluteus larva. Although some nuclear housekeeping or structural proteins might be expected to exhibit such stability, the combination of tissue specificity and stability would also be appropriate for regulatory molecules involved in determination of cell fate. Third, most tissue-specific nuclear proteins whose synthesis does not begin by 16-cell stage are much less stable. Rapid turnover might be expected for regulatory proteins in cells whose fate is not irreversibly fixed (see Davidson (1989) for further discussion).

Determination and differentiation of the ectoderm

Why study ectoderm?

Until a few years ago, the cells of the sea urchin embryo that received the most attention were the micromere–primary mesenchyme lineage. This was because classical experiments demonstrated the inductive powers of micromeres as well as their ability to give rise to skeletogenic mesenchyme either in various abnormal chimeras (reviewed by Hörstadius, 1973), or when isolated in culture (Okazaki, 1975). A number of primary mesenchyme cell-specific mRNAs have been identified and isolated in recombinant clones recently, but these messages begin to accumulate only a few hours before primary mesenchyme cells ingress into the blastocoel (Benson *et al.*, 1987; Harkey *et al.*, 1988). Despite much effort (Rodgers & Gross, 1978; Tufaro & Brandhorst, 1979; Ernst *et al.*, 1980), no evidence has yet been found for mRNAs that are highly enriched in micromeres, and the maternal molecules responsible for their determination have yet to be identified. In contrast, classical blastomere recombination experiments also show that the fate of different regions of the ectoderm is fixed only gradually during the blastula period. In this section we discuss our current knowledge of events of gene expression during ectoderm differentiation. These data lead to the somewhat paradoxical observation that although ectoderm cells are irreversibly determined much later than are the micromeres, apparently they begin to accumulate tissue-specific mRNAs somewhat earlier, even while their fate remains subject to experimental manipulation.

There are both intellectual and practical reasons for studying ectoderm differentiation. First, classical experiments have shown that differentiation of ectoderm to form oral and aboral regions requires inductive signals from the vegetal pole (reviewed by Hörstadius, 1973). Molecular measures of the ability of animal pole cells to differentiate – to express aboral or oral ectoderm-specific mRNAs or antigens – provide assays for the inductive signals from vegetal cells. Second, specification of oral and aboral regions of the ectoderm reflects the underlying determination of the oral–aboral axis of the embryo. (The oral–aboral axis was first designated by Cameron et al. (1987) and corresponds closely, but not precisely, to the 'dorsal–ventral' axis of the older literature). Long before there is any histological sign of bilateral symmetry, an invisible boundary delineates oral and aboral territories of tissue-specific gene expression in the ectoderm of the early blastula. The process(es) by which the axis that these mRNA spatial patterns reveal is initially specified, is transiently maintained subject to alteration during cleavage and early blastula stages, and is ultimately irreversibly fixed, are all poorly understood at present. Tissue-specific mRNAs provide useful probes for studying these mechanisms. Third, the decision between oral and aboral ectoderm is a decision between two quite different fates. Aboral ectoderm cells withdraw from cell division during late blastula stage and differentiate as a histologically rather uniform set of flattened epithelial cells in the unfed pluteus. Cells in the oral region continue to divide especially in areas that will form the ciliated bands, and give rise to several distinct cell types. Practical advantages of studying ectoderm development are that it comprises a major fraction of the embryo, oral and aboral regions each containing about one-third of the volume of egg cytoplasm; and that ectoderm cells can be separated from endoderm + mesenchyme cells and, to a lesser extent, aboral ectoderm can be partially purified at pluteus stage (McClay, 1986). In the remainder of this section we focus on gene expression in the ectoderm.

Temporal and spatial patterns of gene expression in oral and aboral ectoderm

Distinct, although overlapping, phases of gene expression have been identified in the ectoderm, as presented diagrammatically in Figures 4 and 5. While the messages examined to date permit some initial generalisations about patterns of gene expression at different stages of embryogenesis, it should be pointed out that the mRNAs discussed have been identified in different selective screens. Therefore, it is premature to make generalisations about the relative numbers of mRNAs in the different classes outlined below.

Early blastomeres of presumptive ectoderm initially share a pattern of gene expression characteristic of all blastomeres during early cleavage. This includes mRNAs probably encoding proteins required for rapid cell division. Maternal histone mRNA released from the pronucleus distributes to all blastomeres (DeLeon *et al.*, 1983), and all blastomeres participate in the massive accumulation of histone mRNAs during early cleavage. Messages encoding two other abundant maternal mRNAs of as yet unidentified function (SpMW5 and SpMW9; unpublished observations of L. Angerer and M. Winkler) are also uniformly distributed in the egg cytoplasm and early embryo (Fig. 4(a)). Although these non-histone mRNAs also may be transcribed after fertilisation, the major feature of their expression is continued decrease in abundance until, when the rate of cell division slows markedly at blastula stage, their abundance is much decreased. We have not detected inhomogeneities in the distribution of any of these mRNAs in cleavage stage embryos.

Towards the end of cleavage (between 9 and 12 h, or around the seventh cleavage) sets of mRNAs begin to appear that ultimately distinguish ectoderm from territories at the vegetal pole. One of these sets is the VEB messages discussed above, all of which appear to be expressed uniformly in presumptive ectoderm, but various members of which may be absent from different blastomeres at the vegetal pole (Fig. 4(b)). A second spatially related set of mRNAs includes several that initially accumulate uniformly in the embryo by 12 h but then decay selectively in vegetal pole lineages. These include messages encoding β-tubulin mRNAs and mRNA for a secreted cilia-associated protein, Spec3 (Fig. 4(c)), whose mRNA abundance is coordinated with that of β-tubulin message (Eldon *et al.*, 1987; Harlow & Nemer, 1987). A third group includes several mRNAs that show reciprocal patterns, being expressed at higher levels at the vegetal pole. Message encoding cytoplasmic actin CyI initially accumulates uniformly, and at blastula stage is at lower concentration in the ectoderm than in cells of the vegetal plate (Fig. 4(d)). Several other mRNAs appear somewhat later (around 20 h, or early mesenchyme blastula) and are restricted to primary and presumptive secondary mesenchyme cells (actin CyIIa, Cox *et al.*, 1986; collagen, Angerer *et al.*, 1988), or exclusively in primary mesenchyme cells (Benson *et al.*, 1987; Harkey *et al.*, 1988) but never accumulate in ectoderm cells (Fig. 4(f)).

Initial events that distinguish oral and aboral ectoderm also include both differential down-regulation and specific accumulation of individual

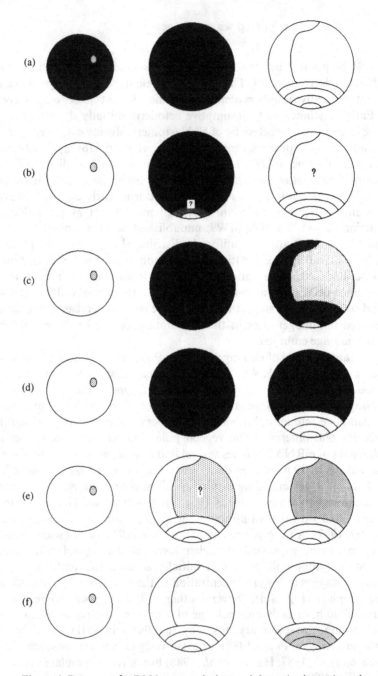

Figure 4. Patterns of mRNA accumulation and decay in the early embryo. The stages illustrated are unfertilised egg, 9–12 h embryo (about 64–150 cell) and 18–20 h embryo (hatching blastula). Levels of shading roughly indicate relative concentration of each individual mRNA in different

mRNAs. The mRNAs listed in Figures 4 and 5 that first accumulate uniformly in the embryo ultimately have different abundances in aboral and oral ectoderm cells. These spatial differences vary in extent, differ in timing, and reflect different features of the aboral and oral ectoderm cell phenotypes. One feature is the difference in cell division in these two territories discussed above. Cytoplasmic actin CyI is maintained in cells that continue to divide, including oral ectoderm, while decaying from those that do not, including aboral ectoderm (Cox et al., 1986) (Fig. 5(a)). Differences in β-tubulin and Spec3 mRNAs are related to production of cilia. All ectoderm cells initially are ciliated, and the difference in content of these mRNAs in oral and aboral regions becomes pronounced only in pluteus stage when these mRNAs accumulate especially in oral ectoderm cells of the ciliated band (Fig. 5(b)).

Some of the most interesting differences in patterns of message accumulation in oral and aboral ectoderm involve expression of tissue-specific genes. Relatively abundant aboral ectoderm mRNAs must encode some of the major proteins responsible for this cell phenotype. These cells have an important rôle in creating the shape of the pluteus larva by expansion of the aboral side, which is mediated by a change in shape of aboral ectoderm cells from cuboidal to flattened epithelial. It has been suggested that proteins

regions of the embryo and at different stages; comparisons are not intended to be made among different mRNAs. (a) Two members of the set of abundant maternal mRNAs that are expressed only during cleavage, SpMW5 and SpMW9 (Grainger et al., 1986). (b) The very early blastula set of messages described in the text (Reynolds et al., in preparation). The 9–12 h stage is shaded to indicate that all VEB mRNAs initially accumulate in ectoderm, and different ones may label different regions of the vegetal pole, but that has not yet been confirmed (indicated by ?). Patterns at 18–20 h are complex, and have not yet been sufficiently defined to illustrate here. (c) Cytoplasmic actin Cy1 (Cox et al., 1986). (d) Spec3 and a β tubulin probe that reacts with more than one mRNA (Eldon et al., 1987). Constitutively expressed metallothionein shows a similar pattern at early stages, but does not label cells in the primary mesenchyme lineage (Angerer et al., 1986). (e) Aboral ectoderm-specific mRNAs CyIIIa and Spec1 appear in the embryo by 9–12 h (Angerer et al., 1989) but their spatial distribution at these early stages is not known (indicated by ?). (f) Messages restricted to vegetal pole lineages include collagen (primary and secondary mesenchyme, Angerer et al., 1988); cytoplasmic actin CyIIa (mainly in secondary mesenchyme, and transiently in primary mesenchyme, Cox et al., 1986); and SM50 (primary mesenchyme-specific; Benson et al., 1987). No messages have been found to be expressed specifically in presumptive endoderm at these early stages.

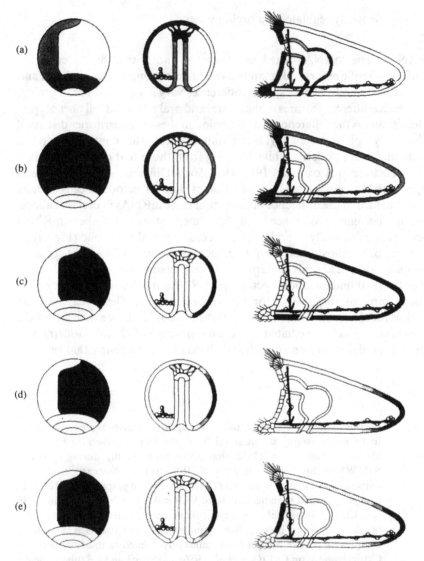

Figure 5. Patterns of mRNA accumulation and decay during ectoderm differentiation. Stages illustrated are mesenchyme blastula (about 24 h), gastrula (about 40–48 h) and pluteus (about 72–84 h). (a) Actin CyI (Cox *et al.*, 1986) accumulates in oral ectoderm, and some mesenchyme cells, and decays in aboral ectoderm. (b) Spec 3 (Eldon *et al.*, 1987) is ectoderm-specific at these stages, and in the late pluteus is especially abundant in cells of the ciliary band. (c) Spec1 (Lynn *et al.*, 1983), actin CyIIIa (Cox *et al.*, 1986) and SpARS (Yang *et al.*, 1989a) mRNAs accumulate specifically in aboral ectoderm. (d) SpHbox1 mRNA is progressively restricted within the aboral ectoderm (Angerer *et al.*, 1989). (e) The distribution of SpEGF2 is similar to that of SpHBox1, but it also accumulates in epithelial oral ectoderm cells around the mouth (Yang *et al.*, 1989b).

encoded by two small gene families, cytoplasmic actins CyIIIa and CyIIIb and the Spec1 and 2 proteins (which belong to the calmodulin/troponinC/myosin light chain group of calcium-binding proteins; reviewed by Hardin *et al.*, 1988) may function either in this shape change, or in the converse change that takes place much later in the ectoderm of the full-grown pluteus upon metamorphosis (Carpenter *et al.*, 1984). Spec1 and actin CyIIIa mRNAs are the earliest aboral ectoderm-specific messages detected in the embryo: *in situ* hybridisation shows that they mark presumptive aboral ectoderm as early as 18 h, or at about 300-cell stage. *In vitro* nuclear transcription assays (Hickey *et al.*, 1988) and RNase protection assays of whole embryo RNA (Angerer *et al.*, 1989) indicate that transcription of these genes begins between 9 and 12 h, although we have not yet documented whether RNA accumulation is tissue-specific at this early stage.

Other aboral ectoderm-specific mRNAs appear in sequence during differentiation of this tissue. Spec1, Spec2a/2c and Spec2d mRNAs accumulate sequentially, being activated at early blastula, mesenchyme blastula, and gastrula stages, respectively (Hardin *et al.*, 1988). A third major mRNA, SpARS, also begins to accumulate in aboral ectoderm at blastula stage, clearly after actin CyIIIa and Spec1 messages (Yang *et al.*, 1989a). SpARS message encodes an arylsulphatase whose enzymatic function can be demonstrated easily, but whose natural substrate(s) and function in the sea urchin embryo remain unknown (Rapraeger & Epel, 1981; Sasaki *et al.*, 1987). SpHbox1 also begins to accumulate at the same time as SpARS mRNA, as discussed below.

One might anticipate that genes expressed specifically in oral ectoderm would also be activated at about the same time as are the aboral ectoderm-specific genes. Although we have analysed spatial patterns for about 40 different mRNAs, none is expressed *only* in oral ectoderm or one of its subregions. This is particularly surprising for cells of the ciliated band, which have distinctive morphology. A frequently occurring pattern of expression (8 of 14 clones, randomly selected from a gastrula stage cDNA library, that encode abundant, developmentally up-regulated mRNAs) is in oral ectoderm, especially ciliated band, and endoderm in the later embryo (P. Kingsley, in preparation). This correlates loosely with continued cell division, but not in a way that corresponds to division *per se*, because such probes do not label cleavage-stage embryos and tend to label entire regions that contain dividing cells, whereas probes for histone mRNAs label patches of cells with residual mitotic synchrony (Cox *et al.*, 1984).

Because of the differences in sensitivity of various assays used to measure first appearance, or disappearance, of different mRNAs, the relative timing of these events is not precisely known. However, it is clear that down-regulation of some mRNAs in the vegetal pole, activation of ectoderm-specific and of the first aboral ectoderm-specific mRNAs all take place within a few hours of each other beginning around the sixth or seventh cleavage. Since oral and aboral lineages separate, at least in the animal hemisphere, at fifth cleavage (discussed by Davidson, 1989) and presumptive endoderm and mesenchyme separate from the ectodermal contribution of the vegetal blastomeres at sixth cleavage, accumulation of ectoderm and aboral ectoderm-specific mRNAs thus begins within a few hours, and no more than two cell divisions, after the corresponding lineages are separated by early cleavage (reviewed also by Davidson, 1989). Since the fates of ectoderm cells remain subject to experimental manipulation during this time, it follows that these initial commitments are alterable, and that interactions among cells are important in maintaining these commitments in normal embryos. We shall return to this point shortly.

mRNAs with unusual spatial distributions in the aboral ectoderm – Hbox1 and SpEGF2

Aboral ectoderm consists of a histologically uniform cell type. This homogeneity is echoed at the molecular level by the observation that probes for a number of different mRNAs all label aboral ectoderm uniformly, as originally observed for Spec1 (Lynn *et al.*, 1983). This is true at all stages of development from blastula to pluteus, and for both aboral ectoderm-specific messages as well as those more broadly expressed. Furthermore, modulations of abundance, both developmental (Cox *et al.*, 1986; Hardin *et al.*, 1988) and physiological (e.g. induction of metallothionein mRNA, Angerer *et al.*, 1985) occur uniformly throughout this tissue, so that a discrete boundary of message expression rather than an oral–aboral gradient is always observed.

The first indication of regional differences within aboral ectoderm came from studies undertaken in collaboration with Drs T. Humphreys and G. Dolecki (Angerer *et al.*, 1989) in which we observed an unusual and conserved pattern of distribution of an mRNA encoded by a gene, Hbox1, containing a homeobox sequence. This mRNA appears just before hatching blastula stage and initially accumulates in most, if not all, cells of the presumptive aboral ectoderm. Soon, however, it begins to disappear in a wave that progresses from the oral–aboral border, and gradually withdraws toward the pointed vertex on the aboral side of the pluteus. Hbox1 mRNA is confined to a region of the pluteus closely corresponding to the aboral ectoderm derived from a single blastomere of the 8-cell embryo (the VA blastomere; Cameron *et al.*, 1987) (Figs 5(d) and 6(a)). The direction of this progressive restriction does not correspond precisely to that of the oral–aboral embryonic axis (see Angerer *et al.*, 1989) and cells in aboral ectoderm of plutei that retain Hbox1 mRNA are not histologically definable. The facts

Spatially regulated sea urchin genes

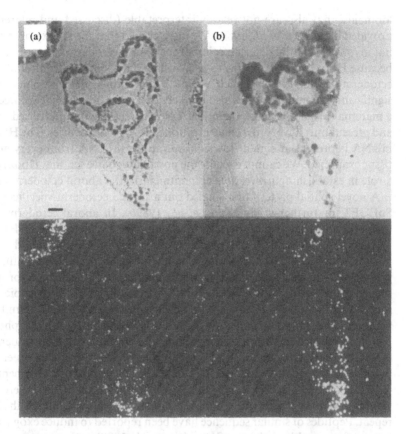

Figure 6. Two mRNAs expressed in novel patterns within the aboral ectoderm. (a) SpHbox1 at pluteus stage, showing labelling restricted to aboral ectoderm near the vertex. (b) SpEGF2 mRNA at pluteus stage is expressed in a pattern similar to that of SpHbox1, but also accumulates in epithelial oral ectoderm cells around the mouth. *In situ* hybridisation was carried out as described in references cited in the text, using SpHbox1 and SpEGF1 probes labelled with ^{35}S (5×10^8 dpm/μg) or ^{3}H (1×10^8 dpm/μg), respectively. Exposure times were 14 and 42 days, respectively. The bar represents 10 μm.

that Hbox1 protein is presumably a DNA binding regulatory protein and that the message distribution reflects the direction of elongation of the aboral side of the embryo imply the presence of spatial information that we cannot yet interpret regulating processes of which we are unaware. Similarly, messages encoding homeoproteins in *Xenopus* and mammalian embryos have distributions that ignore histological borders and define spatial 'domains' that also are not readily interpretable (discussed and references provided in Angerer et al., 1989). We have argued that the Hbox1 protein is not involved in initial determination of the aboral ectoderm cell type *per se*, because Hbox1 mRNA begins to accumulate only some hours after aboral ectoderm-specific mRNAs such as Spec1 and actin CyIIIa have achieved significant accumulation. It is also unlikely that early function is executed by a maternal protein because the mRNA is undetectable in unfertilised eggs, and present only at very low concentration in RNA from ovaries. The Hbox1 mRNA is present at similar low concentrations in RNA from every adult tissue examined. We cannot exclude the possibility, however, that Hbox1 has a rôle in establishing *irreversible* commitment to the aboral ectoderm fate.

A novel spatial pattern first singled out a second ectoderm-specific gene, SpEGF2 for further study (Yang et al., 1989b). In aboral ectoderm, the temporal and spatial patterns of expression are similar to that of Hbox1 message (Figs 5(e) and 6(b)). However, SpEGF2 mRNA is also found in the portion of oral ectoderm consisting of epithelial cells around the mouth, but probably not in cells of the ciliary band. Examination of the deduced protein sequence showed that the majority of the open reading frame is occupied by four tandem repeats of a sequence that clearly belongs to the EGF family of peptides. The amino terminus of SpEGF2 is occupied by a hydrophobic region suitable to serve as a leader peptide, in common with all members of the EGF family. Only about 30 additional amino acids are found between the putative leader peptide and the EGF repeats, suggesting that this peptide may function as a secreted growth factor. Consistent with this, there are several potential proteolytic processing signals, two of which flank the last repeat. Peptides of similar sequence have been reported to induce exogastrulation of sea urchin embryos (Suyemitsu et al., 1989) and we suspect that SpEGF2 is secreted into the blastocoel and may have a rôle in the differentiation of cells during skeletogenesis and/or gastrulation.

SpARS protein is extracellular (Rapraeger & Epel, 1981) and SpEGF2 polypeptides are also likely to function as secreted factors. The messages encoding both proteins are clearly concentrated on the basal sides of cells (Yang et al., 1989a; Yang et al., unpublished observations), which is also the location of the Golgi complex. These results suggest that intracellular *in situ* hybridisation patterns may provide clues about protein trafficking.

The requirement for cell contacts in ectoderm differentiation

The above analysis of temporal and spatial patterns of expression of individual genes provides a time line of quantitatable molecular events that can be used to measure the extent to which embryos subjected to experimental manipulation can commit to patterns of mRNA accumulation and decay typical of those in the normal embryo. Because accumulation and down-regulation of mRNAs precede observable morphological differentiation by a number of hours, these molecular assays are especially useful for analysing early events *per se*, rather than requiring inference from subsequent histological differentiation. Furthermore, expression of individual genes can be assayed separately, avoiding the very stringent requirement for production of a complete morphological structure. In a first approach toward using molecular assays to investigate more precisely the requirement for interactions among blastomeres, Hurley *et al.* (1989) analysed the levels of a number of different mRNAs in the progeny of blastomeres separated at 16-cell stage and maintained separate in Ca^{2+}-free sea water. The results of these measurements support a model in which early events of mRNA accumulation and decay, presumably under the control of maternal regulatory molecules, are carried out rather normally. The cells continue to divide at nearly normal rates during the period analysed, up to 24 h. mRNAs that accumulate early and uniformly in the embryo and in the ectoderm, such as actins CyI and CyIIb, also accumulate in dissociated cells. Messages that are largely maternal, such as SpMW5 and SpMW9, also decay at approximately normal rates. Thus, many of the early events of mRNA accumulation and decay characteristic of all ectoderm appear to be autonomously controlled in separated cells, as would be expected if they required only maternal regulators. In contrast, modulations of mRNA abundance that reflect specification of oral and aboral ectoderm are significantly suppressed. For example, accumulation of Spec1 and actin CyIIIa mRNAs is, on average, reduced by 7- to 9-fold. Furthermore, analysis of Spec1 mRNA content in separated cells by *in situ* hybridisation showed that its accumulation is suppressed in all cells; residual expression is not due to a small subset of cells that contain near-normal levels of Spec1 message. The fact that other mRNAs (actins CyI and CyIIb) over-accumulate markedly compared to controls at early blastula stage is consistent with the idea that the normal down-regulation of these mRNAs in presumptive aboral ectoderm of intact embryos also does not take place in separated cells.

Primary mesenchyme cells have been shown to differentiate independently from micromeres separated at the 16-cell stage. In agreement with this, several mRNAs expressed only in primary and secondary mesenchyme of the normal embryo accumulate to more normal levels. Hurley *et al.* (1989) showed that a collagen message expressed in primary and secondary mesenchyme accumulates to one-third of normal levels, and in a similar set of experiments, Stephens *et al.* (1989) showed that a message, Sm50, encoding the major spicule matrix protein secreted by primary mesenchyme cells accumulates to about one-half of normal levels. Hurley *et al.* (1989) also found evidence for negative interactions among vegetal pole cells. Actin CyIIa mRNA is normally expressed in the early embryo only in presumptive secondary and, to a lesser extent, primary mesenchyme cells. However, again using *in situ* hybridisation we observed a large increase in the number of cells accumulating actin CyIIa mRNA in separated cell populations. This result suggests that interactions among cells in the intact embryo normally suppress accumulation of CyIIa mRNA in some cells. These molecular data parallel the recent demonstration by Ettensohn & McClay (1988) that primary mesenchyme cells suppress differentiation of secondary mesenchyme cells along a primary mesenchyme cell pathway.

These early experiments demonstrate the usefulness of molecular markers in analysis of the results of experimental embryology of the sea urchin embryo. Future experiments will involve recombining different blastomeres in different combinations. Quantity limitations, as well as the greater information afforded by spatial information, will make *in situ* hybridisation a useful method for analysing gene expression in such chimeras.

Concluding observations on patterns of gene expression in the sea urchin embryo

Results from *in situ* hybridisation have provided some answers to questions initially set out. Of all the different mRNAs we have analysed, the only ones that are uniformly distributed in the embryo are the abundant maternal mRNAs expressed only during cleavage, which probably function in rapid cell division (Grainger *et al.*, 1986). All other messages show varying degrees of spatial restriction at some, and usually at most, different stages of development and the spatial distribution of many change during embryogenesis, most frequently by progressive restriction. This is true even for abundant messages which, as a group, might have been expected to be abundant because they encode ubiquitous 'housekeeping' proteins. The examples discussed above, and 14 additional abundant mRNAs, identify three frequently encountered mRNA distributions in later stages of development: aboral ectoderm, oral ectoderm (especially ciliary band) + endoderm, and primary mesenchyme. Analysis of representatives of the major group of abundant mRNAs expressed at approximately constant levels during development (Flytzanis *et al.*, 1982), as well as of rarer messages is currently under way to elucidate dominant spatial patterns of mRNA accumulation (Kingsley *et al.*, in preparation). Several mRNAs examined to date (e.g. muscle actin, Cox *et al.*, 1986; SpHbox1 at later stages, Angerer *et al.*, 1989; Spec2d, Hardin *et al.*, 1988) that are members of the rare mRNA class in the whole embryo have also been found to be tissue-specific and, in two cases, restricted to relatively small regions of embryos. We think it is likely that although most mRNAs are present throughout development, many will be subject to spatial regulation.

Several sets of co-ordinately expressed mRNAs have been identified in sea urchin embryos. The first of these was identified by using RNAs extracted from isolated primary mesenchyme cells (Benson et al., 1987; Harkey et al., 1988). In situ hybridisation analyses have added a number of aboral ectoderm-specific mRNAs, which share patterns of spatial restriction, but differ in apparent time of transcriptional activation. There are at least ten different messages in the oral ectoderm–endoderm spatial set, but their temporal patterns have not yet been analysed in detail. Finally, different mRNAs of the very early blastula set show tight temporal co-ordination, and identify at least one early spatial pattern. Regulatory regions of these genes may provide the starting points from which mechanisms of determination along the animal–vegetal and oral–aboral axes are eventually unravelled. In our view the critical contributions of in situ hybridisation to the study of sea urchin development have been twofold: 1) the identification of these tissue-specific genes as tools for probing developmental and gene regulatory mechanisms, and 2) the demonstration that lineage- and tissue-specific gene expression is underway even before cleavage is completed. It is amusing that this molecular pattern is established at a stage of sea urchin development which morphologically is the most undistinguished, being somewhere between late cleavage and the very early blastula stage.

Acknowledgments

Work in our laboratory is supported by a grant from the National Institutes of Health (GM25553) and by a Research Career Development Award to R. C. A. from the Public Health Service (HD602). The laboratories of Eric Davidson, Tom Humphreys and Greg Dolecki, Larry Kedes, Bill Klein, Martin Nemer, Robert Simpson, Eric Weinberg and Matt Winkler developed many of the cloned probes described here.

References

Angerer, L. M. & Angerer, R. C. (1981). Detection of poly (A)$^+$ RNA in sea urchin eggs and embryos by quantitative in situ hybridization. *Nucleic Acids Research*, **9**, 2819–40.

Angerer, L. M., Chambers, S. A., Yang, Q., Venkatesan, M., Angerer, R. C. & Simpson, R. T. (1988). Expression of a collagen gene in mesenchyme lineages of the *S. purpuratus* embryo. *Genes & Development*, **2**, 239–46.

Angerer, L. M., Cox, K. H. & Angerer, R. C. (1987a). Identification of tissue-specific gene expression by in situ hybridization. In *Methods in Enzymology*, vol. **152**, *Guide to Molecular Cloning Techniques* (eds

Berger, S. & Kimmel, A.) pp. 649–61, New York: Academic Press, Inc.

Angerer, L. M., DeLeon, D. V., Maxson, R., Kedes, L. H., Kaumeyer, J., Weinberg, E. & Angerer, R. C. (1985). Simultaneous expression of early and late histone genes in individual cells of developing sea urchin embryos. *Developmental Biology*, **112**, 157–66.

Angerer, L. M., Dolecki, G. J., Gagnon, M. L., Lum, R., Wang, G., Yang, Q., Humphreys, T. & Angerer, R. C. (1989). Progressively restricted expression of a homeo box gene within the aboral ectoderm of developing sea urchin embryos. *Genes & Development*, **3**, 370–83.

Angerer, L. M., Kawczynski, G., Wilkinson, D., Nemer, M. & Angerer, R. C. (1986). Spatial patterns of expression of metallothionein mRNA in sea urchin embryos. *Developmental Biology*, **116**, 543–7.

Angerer, L. M., Stoler, M. H. & Angerer, R. C. (1987b). In situ hybridization with RNA probes – an annotated recipe. In *In situ hybridization – Applications to Neurobiology* (eds Valentino, K. L., Eberwine, J. H. & Barchas, J. D.) pp. 71–96, Oxford: Oxford University Press.

Angerer, R. C., Cox, K. H. & Angerer, L. M. (1987c). *In situ* hybridization to cellular RNAs. *Genetic Engineering*, **7** 43–65.

Angerer, R. C. & Davidson, E. H. (1984). Molecular indices of cell lineage specification in the sea urchin embryo. *Science*, **226**, 1153–60.

Benson, S. C., Sucov, H., Stephens, L., Davidson, E. & Wilt, F. (1987). A lineage-specific gene encoding a major matrix protein of the sea urchin embryo spicule. I. Authentication of the cloned gene and its developmental expression. *Developmental Biology*, **120**, 499–506.

Brahic, M. & Haase, A. T. (1978). Detection of viral sequences of low reiteration frequency by *in situ* hybridization. *Proceedings of the National Academy of Sciences, USA*, **75**, 6125–9.

Bruskin, A., Tyner, A. L., Wells, D. E., Showman, R. M. & Klein, W. H. (1982). Accumulation in embryogenesis of five mRNAs enriched in the ectoderm of the sea urchin pluteus. *Developmental Biology*, **87**, 308–18.

Butler, E. & Chamberlin, M. J. (1982). Bacteriophage SP6–RNA polymerase I. Isolation and characterization of the enzyme. *Journal of Biological Chemistry*, **257**, 5772–8.

Calzone, F. J., Thèzè, N., Thiebaud, P., Hill, R. L., Britten, R. J. & Davidson, E. H. (1988). Developmental appearance of factors that bind specifically to *cis*-regulatory sequences of a gene expressed in the sea urchin embryo. *Genes & Development*, **2**, 1074–88.

Cameron, R. A., Hough-Evans, B. R., Britten, R. J. & Davidson, E. H. (1987). Lineage and fate of each blastomere of the eight-cell sea urchin embryo. *Genes & Development*, **1**, 75–85.

Carpenter, C. D., Bruskin, A. M., Hardin, P. E., Keast, M. J., Anstrom, J., Tyner, A. L., Brandhorst, B. P. & Klein, W. H. (1984). Novel proteins belonging to the troponin C superfamily are encoded by a set of mRNAs in sea urchin embryos. *Cell*, **36**, 663–71.

Cox, K. H., Angerer, L. M., Lee, J. J., Davidson, E. H. & Angerer, R. C. (1986). Cell lineage-specific programs of expression of multiple actin genes during sea urchin embryogeneiss. *Journal of Molecular Biology*, **188**, 159–72.

Cox, K. H., DeLeon, D. V., Angerer, L. M. & Angerer, R. C. (1984). Detection of mRNAs in sea urchin embryos by *in situ* hybridization using asymmetric RNA probes. *Developmental Biology*, **101**, 485–502.

Davidson, E. H. (1976). *Gene Activity in Early Development*, 2nd edn. New York: Academic Press.

Davidson, E. H. (1986). *Gene Activity in Early Development*, 3rd edn. Orlando, Florida: Academic Press.

Davidson, E. H. (1989). Lineage specific gene expression and the regulative capacities of the sea urchin embryo: A proposed mechanism. *Development*, **105**, 421–5.

DeLeon, D. V., Cox, K. H., Angerer, L. M. & Angerer, R. C. (1983). Most early variant histone mRNA is contained in the pronucleus of sea urchin eggs. *Developmental Biology*, **100**, 197–206.

Eldon, E. D., Angerer, L. M., Angerer, R. C. & Klein, W. H. (1987). Spec3: Embryonic expression of a sea urchin gene whose product is involved in ectodermal ciliogenesis. *Genes & Development*, **1**, 1280–92.

Ettenshohn, C. A. & McClay, D. R. (1988). Cell lineage conversion in the sea urchin embryo. *Developmental Biology*, **125**, 396–409.

Ernst, S. G., Hough-Evans, B. R., Britten, R. J. & Davidson, E. H. (1980). Limited complexity of the RNA in micromeres of 16 cell sea urchin embryos. *Developmental Biology*, **79**, 119–27.

Flytzanis, C. N., Brandhorst, B. P., Britten, R. J. & Davidson, E. H. (1982). Developmental patterns of cytoplasmic transcript prevalence in sea urchin embryos. *Developmental Biology*, **91**, 27–35.

Franks, R. R., Hough-Evans, B. R., Britten, R. J. & Davidson, E. H. (1988*a*). Direct introduction of cloned DNA into the sea urchin zygote nucleus, and fate of injected DNA. *Development*, **102**, 287–99.

Franks, R. R., Hough-Evans, B. R., Britten, R. J. & Davidson, E. H. (1988*b*). Spatially deranged though temporally correct expression of a *Strongylocentrotus purpuratus* actin gene fusion in transgenic embryos of a different sea urchin family. *Genes & Development*, **2**, 1–12.

Galau, G. A., Lipson, E. D., Britten, R. J. & Davidson, E. H. (1977). Synthesis and turnover of polysomal mRNAs in sea urchin embryos. *Cell*, **10**, 415–32.

Grainger, J. L., von Brunn, A. & Winkler, M. M. (1986). Transient synthesis of a specific set of proteins during the rapid cleavage phase of sea urchin development. *Developmental Biology*, **114**, 403–15.

Hafen, E., Levine, M., Garber, R. L. & Gehring, W. J. (1983). An improved in situ hybridization method for the detection of cellular RNAs in *Drosophila* tissue sections and its application for localizing transcripts of the homeotic *Antennapedia* gene complex. *The EMBO Journal*, **2**, 617–23.

Hardin, P. E., Angerer, L. M., Hardin, S.H., Angerer, R. C. & Klein, W. H. (1988). The Spec2 genes of *Strongylocentrotus purpuratus*: Structure and differential expression in embryonic aboral ectoderm cells. *Journal of Molecular Biology*, **202**, 417–31.

Harkey, M. A., Whiteley, H. R. & Whitely, A. H. (1988). Coordinate accumulation of primary mesenchyme-specific transcripts during skeletogenesis in the sea urchin embryo. *Developmental Biology*, **125**, 381–95.

Harlow, P. & Nemer, M. (1987). Coordinate and selective β tubulin gene expression associated with cilium formation in sea urchin embryos. *Genes & Development*, **1**, 1293–304.

Hickey, R. J., Boshar, M. F. & Crain, Jr., W. R. (1988). Transcription of three actin genes and a repeated sequence in isolated nuclei of sea urchin embryos. *Developmental Biology*, **124**, 215–27.

Hörstadius, S. (1973). *Experimental Embryology of Echinoderms*. Oxford: Clarendon Press.

Hough-Evans, B. R., Franks, R. R., Cameron, R. A., Britten, R. J. & Davidson, E. H. (1987). Correct cell type-specific expression of a fusion gene injected into sea urchin eggs. *Developmental Biology*, **121**, 576–9.

Hurley, D., Angerer, L. M. & Angerer, R. C. (1989). Altered expression of spatially regulated embryonic genes in the progeny of separated sea urchin blastomeres. *Development*, **106**, 567–79.

Lawrence, J. B. & Singer, R. H. (1985). Quantitative analysis of *in situ* hybridization methods for the detection of actin gene expression. *Nucleic Acids Research*, **13**, 1777–99.

Lynn, D. A., Angerer, L. M., Bruskin, A. M., Klein, W. H. & Angerer, R. C. (1983). Localization of a family of mRNAs in a single cell type and its precursors in sea urchin embryos. *Proceedings of the National Academy of Sciences, USA*, **80**, 2656–60.

McClay, D. R. (1986). Embryo dissociation, cell isolation and cell reassociation. *Methods in Cell Biology*, **2**, 309–23.

McClay, D. R. & Wessel, G. M. (1985). The surface of the sea urchin embryo at gastrulation: A molecular mosaic. *Trends in Genetics*, **1**, 12–16.

Okazaki, K. (1975). Spicule formation by isolated micromeres of the sea urchin embryo. *American Zoologist*, **15**, 567–81.

Overton, C. & Weinberg, E. S. (1978). Length and sequence heterogeneity of the histone gene repeat unit of the sea urchin, S. purpuratus. *Cell*, **14**, 247–58.

Pehrson, J. R. & Cohen, L. H. (1976). The fate of the small micromeres in sea urchin development. *Developmental Biology*, **113**, 522–6.

Rapraeger, A. C. & Epel, D. (1981). The appearance of an extracellular arylsulfatase during morphogenesis of the sea urchin *Strongylocentrotus purpuratus*. *Developmental Biology*, **88**, 269–78.

Rodgers, W. H. & Gross, P. R. (1978). Inhomogeneous distribution of egg RNA sequences in the early embryo. *Cell*, **14**, 279–88.

Sasaki, H., Akasaka, K., Shimada, H. & Shimada, T. (1987). Purification and characterization of arylsulfatase from sea urchin (*Hemicentrotus pulcherrimus*) embryos. *Comparative Biochemistry & Physiology*, **88B**, 147–52.

Servetnick, M. D. & Wilt, F. H. (1987). Changes in the synthesis and intracellular localization of nuclear proteins during embryogenesis in the sea urchin *Strongylocentrotus purpuratus. Developmental Biology,* **123**, 231–44.

Showman, R. M., Wells, D. E., Anstrom, J., Hursh, D. A. & Raff, R. A. (1982). Message-specific sequestration of maternal histone mRNA in the sea urchin egg. *Proceedings of the National Academy of Sciences, USA,* **79**, 5944–7.

Singer, R. H., Lawrence, J. B., and Rashtchian, R. N. (1987). Toward a rapid and sensitive *in situ* hybridization methodology using isotopic and nonisotopic probes. In In situ *Hybridization – Applications to Neurobiology* (eds Valentino, K. L., Eberwine, J. H. & Barchas, J. D.) pp. 71–96, Oxford: Oxford University Press.

Stephens, L., Kitajima, T. & Wilt, F. (1989) Autonomous expression of tissue-specific genes in dissociated sea urchin embryos. *Development,* **107**, 299–307.

Sucov, H. M., Hough-Evans, B. R., Franks, R. R., Britten, R. J. & Davidson, E. H. (1988). A regulatory domain that directs lineage-specific expression of a skeletal matrix protein gene in the sea urchin embryo. *Genes & Development,* **2**, 1238–50.

Suyemitsu, T., Asami-Yoshizumi, T., Noguchi, S., Tonegawa, Y., & Ishihara, K. (1989). The exogastrula-inducing peptides in embryos of the sea urchin, *Anthocidaris crassispina* – isolation and determination of the primary structure. *Cell Differentiation and Development,* **26**, 53–66.

Tufaro, F. & Brandhorst, B. P. (1979). Similarity of proteins synthesized by isolated blastomeres of early sea urchin embryos. *Developmental Biology,* **71**, 390–7.

Venezky, D. L., Angerer, L. M. & Angerer, R. C. (1981). Accumulation of histone repeat transcripts in the sea urchin egg pronucleus. *Cell,* **24**, 385–91.

Yang, Q., Angerer, L. M., & Angerer, R. C. (1989*a*). Structure and tissue-specific developmental expression of a sea urchin arylsulfatase gene. *Developmental Biology,* **135**, 53–65.

Yang, Q., Angerer, L. M., & Angerer R. C. (1989*b*). Unusual pattern of accumulation of mRNA encoding an E6F-related protein in sea urchin embryos. *Science,* **246**, 806–8.

Alternate probes are hybridised to sections of embryos on a microscope slide. When developed following autoradiography, each slide will yield infor-

P. W. Ingham, A. Hidalgo and A. M. Taylor

Advantages and limitations of *in situ* hybridisation as exemplified by the molecular genetic analysis of *Drosophila* development

Introduction

Over the past five years the application of *in situ* hybridisation to the analysis of gene expression has made an immense contribution to the study of *Drosophila* development (for reviews, see Akam, 1987; Ingham, 1988). The advent of this technique not only allowed the verification of the inferences about normal gene expression based upon classical genetic analysis, but has also facilitated the investigation of regulatory interactions between genes. In addition, in a number of cases, *in situ* analysis has revealed novel rôles for genes not previously predicted by mutational analysis.

In the first part of this paper we outline briefly the methodology and applications of the various *in situ* hybridisation protocols which have been used with this organism. In the second part, we review the recent advances in the analysis of *Drosophila* development and discuss the uses and limitations of the *in situ* technique in the study of genes of unknown biochemical function.

Methodology

Several protocols have been developed for the analysis of transcripts *in situ* (Hafen *et al.*, 1983; Akam, 1983; Ingham, Howard & Ish-Horowicz, 1985; Mahoney & Lengyel, 1987), but probably the most convenient and widely applicable of these employs tissues which have been embedded and sectioned in paraffin wax.

A major advantage of wax-embedded material over frozen tissue is the ease with which it can be handled and sectioned; large amounts of material can be accumulated and stored at different stages of the procedure, either prior to or following embedding. In addition, wax sections can be cut on most microtomes, obviating the need for expensive (and cumbersome) cryostats, an important consideration where money and space are limiting. Finally, considerably less skill is required to cut good wax sections, and the

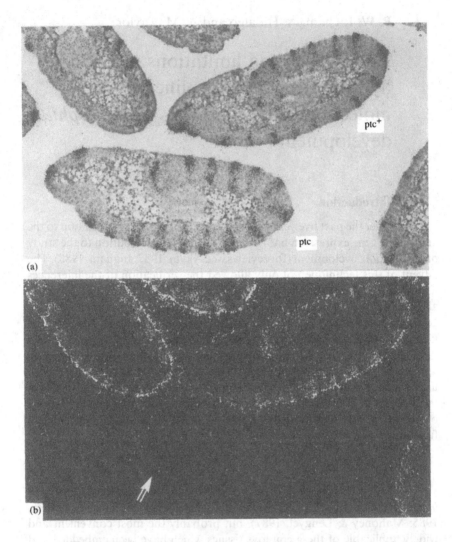

Figure 1. Adjacent sections of *Drosophila* embryos hybridised with probes for the segment polarity genes *wingless* (a) and *patched* (b). The embryos were derived from parents heterozygous for a deletion of the *patched* gene (Guerrero, Hidalgo, Taylor, Whittle & Ingham, in preparation). Probing these embryos with *ptc* probe reveals a homozygous mutant embryo (b, arrowed) when viewed with dark field optics. The same embryo shows an altered pattern of *wingless* expression (a).

morphology is almost invariably superior to that achieved with frozen tissues.

Many of the studies of gene expression in *Drosophila* embryos have involved the simultaneous analysis of two or more genes in the same embryo. In the absence of good methods for differential labelling of probes, these studies have relied upon the hybridisation of adjacent sections with different probes. This approach is useful not only for the comparison of patterns of gene expression, but also as a means of identifying mutant embryos in a mixed population (for example, see Figure 1), an inevitable situation when studying recessive lethal mutations. In the latter case, the mutation used must be a transcript null; sections are then hybridised with probes for the deleted gene and the homozygous embryos identified by the absence of hybridisation.

When such comparisons are not required, an alternative and technically easier method is the hybridisation of whole tissues. Several such protocols have been described (Kornberg *et al.*, 1985; Mahoney & Lengyel, 1987; Baker, 1988). The principal advantage of this technique is that the topography of the tissue is retained thus circumventing the problems of serial reconstruction, and making the description of complex patterns of expression much easier. A disadvantage of this technique is the poor resolution afforded by radioactively labelled probes and compounded by the distortions which often ensue the coating of the hybridised material with photographic emulsion. The use of non-radioactively labelled probes would avoid these problems, and a protocol recently developed by D. Tautz (personal communication) has yielded excellent results with early embryonic stages. Moreover, this technique affords the possibility of simultaneous visualisation of two different transcripts in the same tissue by using different labelled precursors (e.g. biotin or digoxygenin-coupled deoxynucleotides). Different protocols have also employed a variety of probes: double-stranded DNA labelled by nick translation or random priming; single-stranded DNA probes derived from M13 based vectors; and single-stranded RNA probes produced by *in vitro* transcription of sequences cloned in vectors such as *Bluescript* or *pGem*. In our hands, the latter method, first described in detail by Cox *et al.* (1984) has proved the most versatile. Recently, we have made use of *in situ* hybridisation with asymmetric riboprobes to scan a large segment of genomic DNA in order to define the limits and direction of the transcription unit corresponding to the *Drosophila patched* gene (I. Guerrero, A. Hidalgo, A. Taylor, R. Whittle & P. Ingham, in preparation). This method proved extremely rapid and efficient, yielding more information than conventional or reverse Northern analysis from a fraction of the material required by these methods. A generalised scheme of the procedure, which should be applicable to any cloned segment of genomic DNA is presented in Figure 2. The DNA is first digested with a convenient enzyme (such as EcoR1) and subcloned in a transcription vector. The insert is then transcribed separately in each direction with either T3 or T7 polymerase, using mini-prep DNA as template, and [^{35}S] labelled UTP as precursor. Alternate probes are hybridised to sections of embryos on a microscope slide. When developed following autoradiography, each slide will yield information about the direction of transcription and the spatio-temporal distribution of any transcript derived from a particular fragment. Such a technique can be a useful first step in defining the transcriptional limits of a gene, especially when spatial or temporal patterns of expression can be predicted on the basis of mutant phenotypes.

In the following sections we will review how these various *in situ* techniques have been applied to the molecular genetic analysis of *Drosophila* development and consider some implications of the findings of these studies for the analysis of gene expression in other organisms.

The genetic analysis of *Drosophila* development: an historical perspective

Perhaps the most significant contribution so far of *in situ* hybridisation has been in the unravelling of the early events of *Drosophila* embryogenesis. In the 1960s and 1970s much attention focussed on a particular aspect of the development of the fly, namely the genetic control of the diversification of the body parts. This was largely due to the earlier discovery of homoeotic mutations – which cause transformations of one part of the body to another, such as an antenna to leg – and the pioneering analysis by E. B. Lewis (1963; 1978) which established the paradigm for the function of homoeotic genes. The work of Lewis, and subsequently of many other authors (e.g. Garcia-Bellido, 1977; Struhl, 1983), demonstrated that the differentiation of each body segment is under the control of a specific set of homoeotic genes, particularly those organised in the two clusters known as the bithorax and Antennapedia complexes.

The models of homoeotic gene function based on their developmental genetic analysis predicted that they would be expressed continuously from an early stage in development in spatially restricted domains. This prediction was verified in large part by some of the earliest *in situ* hybridisation experiments in *Drosophila* (Levine *et al.*, 1983; Akam, 1983). Transcripts of the *abd-A* gene for instance are found to be restricted to the abdominal segments of the developing embryo, consistent with the effects of *abd-A* mutations which are also limited to these segments (see Fig. 3). It has become clear from these studies, however, that the deployment of homoeotic genes is more complex than had been initially suggested by developmental genetic

In situ analysis of Drosophila development

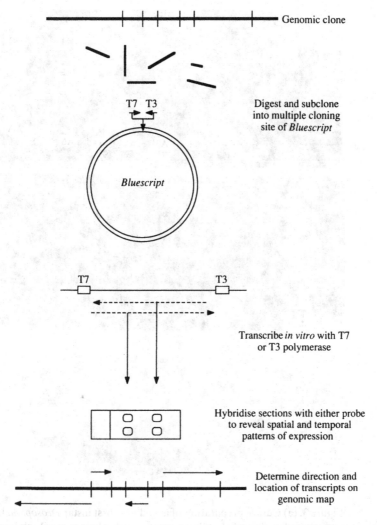

Figure 2. Mapping transcribed regions of genomic DNA by *in situ* hybridisation.

analyses. A particularly striking example is provided by the *Sex combs reduced* (*Scr*) gene, the domains of expression of which change with time both within and between different germ layers (see Martinez-Arias *et al.*, 1987).

Part of the success of the genetic analysis of the homoeotic genes was due to the very nature of their function; different cells adopt different fates according to their particular repertoire of homoeotic gene expression, a combinatorial specification that can be tested by the phenotypic analysis of

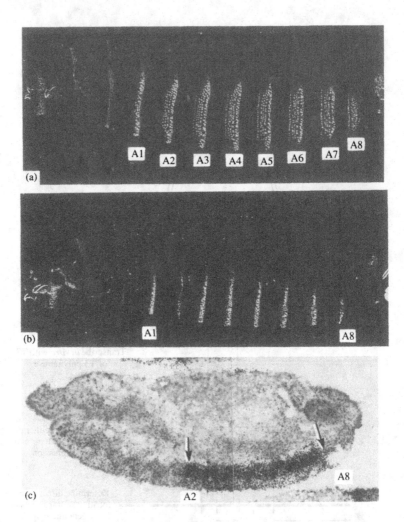

Figure 3. (a) Cuticle preparation of a wild type first instar *Drosophila* larva viewed under dark field illumination. Each thoracic and abdominal segment is characterised by a band of denticles on the ventral surface of the cuticle. Note in particular the A1 denticle belt, which is considerably narrower than those of the other abdominal segments. (b) Similar preparation of a first instar larva homozygous for the homoeotic mutation *abd*-A. This mutation causes the transformation of segments A2–A8 towards A1, as revealed by the change in width of their denticle belts (compare with (a)). (c) Section of a wild type embryo at about 12 h of development which has been hybridised with a probe for *abd*-A transcript. As predicted by the genetic analysis, *abd*-A transcript (indicated by arrows) accumulates in abdominal segments 2–8, but is not expressed more anteriorly.

In situ analysis of *Drosophila* development

animals carrying different combinations of homoeotic mutations (e.g. Struhl, 1983). Homoeosis, however, provides insight into only one aspect of embryonic development, as the homoeotic mutant phenotypes demonstrate. For example, although animals deficient for the *abd-A* gene exhibit a transformation of all their abdominal segments to the first abdominal character (Fig. 3), they still develop the correct *number* of segments, and the organisation of the different germ layer derivatives is unaffected. To discover the genetic basis of the processes underlying these latter aspects of embryonic development, Nusslein-Volhard, Wieschaus and colleagues (Jurgens *et al.*, 1984; Nusslein-Volhard, Wieschaus & Kluding, 1984; Wieschaus, Nusslein-Volhard & Jurgens, 1984) undertook a series of mutagenesis screens designed to identify all the genes required for pattern formation along both the antero-posterior and dorso-ventral axes of the embryo. Their success in this enterprise has provided the basis for much of the study of *Drosophila* development over the past decade, and it is in the analysis of the genes which they identified that *in situ* hybridisation has played a key part.

Identification and analysis of embryonic lethal mutations

Two types of mutation were identified in these screens: those which act maternally and those which act zygotically. In the former case, the mutation affects the progeny of the female which carries it, irrespective of the genotype of the progeny. In the latter case, the mutation affects only the individual which carries it. The existence of these two types of mutation was not in itself surprising, since it had long been considered that when laid, the insect egg is already endowed with positional cues, deposited by the female during oogenesis (Sander, 1976). The maternal effect mutations should therefore identify genes which encode such positional cues, whereas the zygotic mutations identify genes which respond to and elaborate upon them. A good example of the maternal class is the *bicoid* (*bcd*) gene. Developmental analysis of *bcd* mutations suggest that it encodes a determinant required for the development of the anterior half of the embryo (Fronhofer & Nusslein-Volhard, 1986). This hypothesis is strongly supported by the finding that the newly laid egg contains *bcd* RNA localised at its anterior pole (Berleth *et al.*, 1988; see Fig. 4).

Amongst those zygotic mutations which affect pattern formations along the antero-posterior axis, Nusslein-Volhard & Wieschaus (1980) identified three classes in addition to the previously described homoeotics: the gap class and pair rule class, both of which cause the elimination of body segments, and the segment polarity class, which cause the elimination and duplication of different parts of each body segment. Although the phenotypes of the different classes of mutation suggested that the genes might function in the progressive subdivision of the embryo into smaller domains, it was not

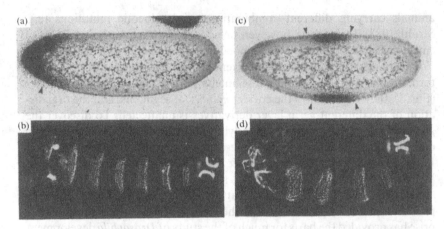

Figure 4. (a) Section of a wild type embryo at the beginning of blastoderm formation, prior to the onset of zygotic gene expression. Transcripts of the *bcd* gene synthesised by the mother and deposited in the egg during oogenesis are clearly detectable at the anterior pole (left) of the embryo (arrow). In the absence of these transcripts, the anterior half of the body fails to form, resulting in the mutant larval phenotype shown in (b). (c) Section of a wild type embryo towards the end of the blastoderm stage. This section has been hybridised with a probe for the zygotically expressed gap gene *Kruppel*. Transcripts of this gene accumulate specifically in the middle region of the embryo (arrows); in the absence of this expression the corresponding part of the body pattern fails to form, as shown in (d).

apparent how conventional developmental genetics could establish precisely the way in which this is accomplished. In contrast to the homoeotic mutations, the terminal phenotype of many of these so-called segmentation mutations is the consequence of cell death rather than transformation (e.g. Ingham *et al.*, 1985; Martinez-Arias, 1985; Lehmann & Nusslein-Volhard, 1987). Such an apparently non-specific defect does not lend itself to the kinds of analysis successfully employed with homoeotic mutations; when combinations of gap and pair rule or pair rule and segment polarity mutations are constructed they generally produce strictly additive phenotypes which give no clue as to potential regulatory interactions between the different classes. The molecular cloning of these genes provided the means to analyse new gene specific phenotypes, namely DNA and RNA probes specific for the transcripts of each gene.

In situ analysis of segmentation gene expression

By using *in situ* hybridisation in combination with genetic analysis it has been possible to elucidate the hierarchy of regulatory interactions which exists between the segmentation genes (see Ingham, 1988). A first discovery, which was by no means a foregone conclusion, is that the genes of the different phenotypic classes are, like the homoeotic genes, expressed in spatially restricted domains which correspond approximately to the regions of the embryo affected by their mutations. Thus gap genes, which when mutant cause the deletion of contiguous regions of the embryonic pattern, are expressed in fairly broad domains in the early embryo (see Fig. 4) whereas pair rule genes, which cause series of smaller repeating deletions are expressed in stripes (Fig. 5). Moreover, these patterns of expression are very transient, evolving and decaying over relatively short periods of time during the initial stages of embryogenesis (Fig. 5). Comparison of the time course of the gap and pair rule patterns of expression support the possibility that the gap genes act to control expression of the pair rules. By studying the pattern of pair rule gene expression in gap gene mutations, it was possible to confirm this inference. For instance, the expression of the pair rule mutation *fushi tarazu*, (*ftz*) is altered in embryos homozygous for the gap mutation *Kruppel* (*Kr*) (Ingham, Ish-Horowicz & Howard, 1986), indicating that *Kr* activity is required for the correct spatial regulations of *ftz* (see Fig. 6). Similar regulatory interactions have also been demonstrated between the gap genes *hunchback* (*hb*) and *knirps* (*kni*) and the pair rule gene *even skipped* (*eve*) as well as *ftz* (Carroll & Scott, 1986; French & Levine, 1987). In addition, interactions within both the gap gene class (Jackle *et al.*, 1986) and the pair rule genes (Howard & Ingham, 1986) were discovered.

These studies clearly showed for the first time that the gap and pair rule genes are involved in the initial specification of cells in the early embryo, at the cellular blastoderm stage of embryogenesis. In contrast to the homoeotic genes, which are required and expressed continuously in spatially restricted domains, the transient expression of the gap and pair rule genes means that they do not function by 'maintaining' specific regions of the body. The cell death and subsequent pattern deficiencies associated with their mutant alleles are thus indirect consequences of the failure of these genes earlier in development; it seems that gap and pair rule mutations have the effect of altering the early fate map of the embryo rather than eliminating defined regions of it (Howard & Ingham, 1986; Ingham *et al.*, 1986). Thus, although there is often a close correspondence between the pattern of expression of a gene and its terminal mutant phenotype, this does not necessarily imply an autonomous effect of the gene in particular groups of cells. Moreover, the extremely transient nature of these early patterns of gene expression make judgements about the critical time and site of expression highly subjective (discussed in Macdonald, Ingham & Struhl, 1986). Whilst the expression domain of *ftz* does correspond, albeit fleetingly, to a segment-wide domain reminiscent of the segment wide deletions caused by *ftz* mutations, the

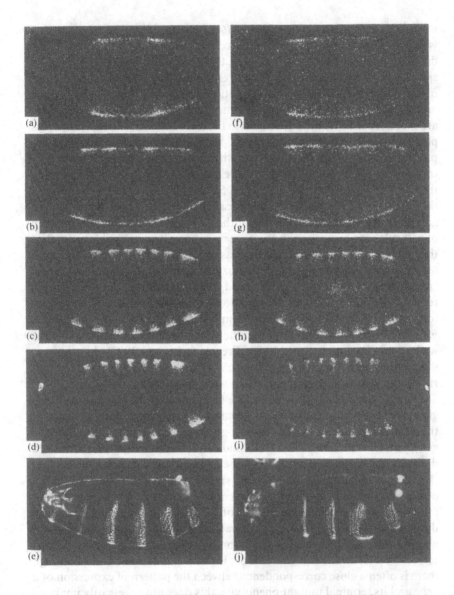

Figure 5. Panels (a)–(d) and (f)–(i) show adjacent sections of successively later blastoderm embryos which have been hybridised with probes for the pair rule genes *fushi tarazu* and *even skipped*. Note how the patterns of seven regularly spaced stripes (c) and (h) evolve from more diffuse and uniform distributions ((a), (b), (f) and (g)) before narrowing further ((d) and (i)). The inference that the seven striped expression pattern of both genes is crucial for normal development is supported by the *ftz* (e) and *eve* (j) mutant phenotypes.

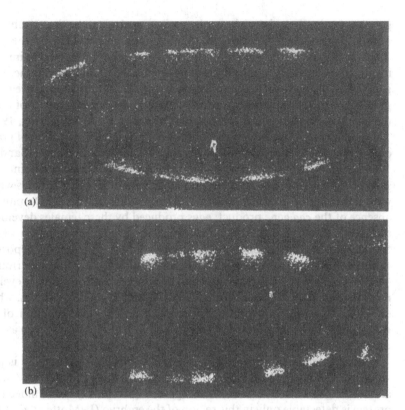

Figure 6. Altered expression patterns of the pair rule genes *hairy* (a) and *ftz* (b) in a blastoderm embryo homozygous for the *Kr* mutation. Both of these genes are normally expressed in a series of seven evenly spaced bands. Note that in both cases the bands are changed in width and distribution, rather than simply eliminated from the normal *Kr* domain.

expression domain of another pair rule gene *paired* (*prd*), does not (Kilcherr et al., 1986). In all cases, pair rule genes are at some point expressed in domains both larger and smaller than those which might be predicted from their mutant phenotypes. These considerations have important implications for the analysis of the expression patterns of genes of unknown function in organisms where genetic analysis is not possible. In particular, it is important to consider that the function of a given gene may not be restricted to those cells which express it. By the same token, expression of a gene may be of no functional significance at all.

Transcription does not necessarily imply function

Several examples of apparently redundant expression patterns have been described in *Drosophila*. In these cases, spatially modulated patterns of gene expression have been shown by genetic analysis to be dispensable for normal development. Perhaps the best example is provided by the gene *caudal* (*cad*). *cad* is unusual amongst *Drosophila* genes in that it was initially identified molecularly, on the basis of its homoeobox homology, rather than genetically. Analysis of *cad* expression by *in situ* hybridisation revealed a gradient of increasing concentration of maternally derived transcript along the antero-posterior axis of the embryo (Mlodzig, Fjose & Gehring, 1985). This finding aroused considerable excitement given the implication of gradients of maternally derived morphogens in early pattern formation (Sanders, 1976). In an elegant set of experiments, Macdonald & Struhl (1986) induced several null mutations of the gene and used these to generate females which were incapable of producing *cad* transcript during oogenesis. Despite the absence of the *cad* gene product, eggs produced by these females developed into perfectly normal flies, demonstrating the graded distribution of the transcript to be of no functional significance. Similarly, maternally deposited transcripts of the gap gene *hb*, which are also transiently gradially distributed (Tautz *et al.*, 1987), have been shown to be unnecessary for normal development (Lehmann & Nusslein-Volhard, 1987). Both of these examples, which are likely to represent evolutionary relics of an ancestral function of the genes, illustrate the importance of the availability of mutations when assessing the significance of the patterns of gene expression.

A third example of apparently redundant transcriptional activity is provided by the homoeotic gene *Scr*. Mutations of *Scr* suggest that it is required in only the gnathal and first thoracic segments (Struhl, 1983) and indeed Scr protein is detectable only in this region of the embryo (LeMotte *et al.*, 1989; see Fig. 7). When it first becomes transcriptionally active at the blastoderm stage of embryogenesis however, it is expressed in seven evenly spaced stripes along the antero-posterior axis (Ingham & Martinez-Arias, 1986), a pattern reminiscent of that of the pair rule genes. Extrapolation of the function of *Scr* from its transcriptional pattern alone, could thus lead to the erroneous conclusion that it functions as a pair rule gene. In fact, the later restricted transcription and translation of the gene along with its mutant phenotype, place it firmly in the homoeotic gene class.

The redeployment of segmentation gene transcripts reveals novel functions for their products

Although mutational analysis of developmental processes is an extremely powerful analytical technique, it also has several shortcomings. Most obviously, it requires that mutations produce a clearly detectable phenotype. This is usually not a problem when considering the development and differentiation of external structures such as those secreted by the insect

In situ analysis of *Drosophila* development 109

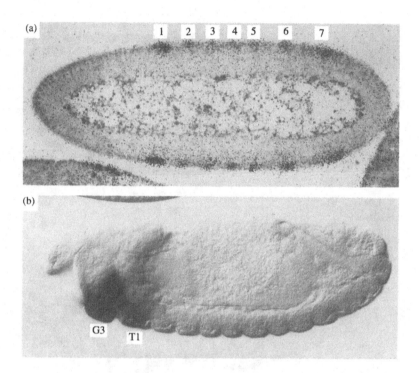

Figure 7. (a) Pattern of expression of the *Scr* gene in a wild type blastoderm embryo. Note the seven domains of expression, reminiscent of the pair rule genes *ftz* and *eve*. (b) Distribution of *Scr* protein in a later stage embryo (at about twelve hours of development) as visualised by staining with a monoclonal antibody raised against an *Scr* fusion protein. The expression of the gene is limited to the third gnathal (G3) and first thoracic (T1) segments, in line with *Scr* mutant phenotype.

epidermis; the phenotypes of mutations in genes required for the development of internal tissues such as the nervous system or mesodermal derivatives, however, are less easy to predict. A second shortcoming relates to the identification of genes with multiple functions. If these functions are temporarily separated, null mutations may only reveal the earliest of them. Both of these limitations have been circumvented by the direct visualisation of segmentation gene expression. Quite unexpectedly, first *ftz* (Carroll & Scott, 1985) and then other genes such as *Kr* (Knipple *et al.*, 1985) and *eve* (Macdonald *et al.*, 1986) were found to be re-expressed at later stages of embryogenesis in patterns which bear no obvious relation to their described mutant phenotypes (see Fig. 8). In particular, they are expressed in specific neurones in each segmental ganglion of the central nervous system in such a

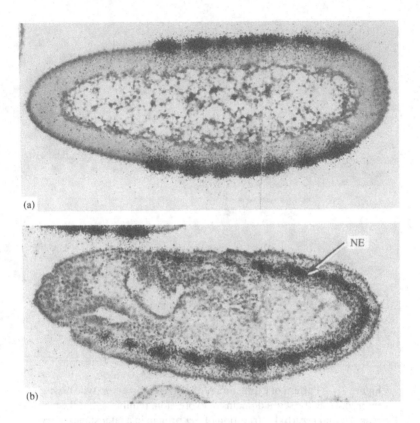

Figure 8. (a) Double segment periodic expression of *ftz* at the cellular blastoderm stage of embryogenesis. (b) After the 'pair rule' pattern shown in (a) has decayed, *ftz* transcription is reinitiated, but now in the neural cells of every segment, as revealed by the almost continuous labelling of the neurectoderm (NE).

precise pattern as to imply a previously undetected role for these genes in neuronal specification. These observations have prompted a re-examination of the function of these genes. By engineering a *ftz* gene lacking the control elements for neuronal transcription, it was found that absence of *ftz* expression results in the transformation of the fate of specific neurones. Similarly, a detailed analysis of a temperature sensitive *eve* allele revealed a requirement for this gene during neuronal differentiation, precisely as predicted from its pattern of transcription (reviewed in Doe & Scott, 1988). Interpretation of the later *Kr* expression is complicated further by the finding of differences between transcriptional and translational domains (Gaul *et al.*, 1987). Whereas *Kr* becomes transcribed throughout the neurogenic region of the embryo, its protein is only detectable in a sub-set of these neuroblasts. The functional significance of this expression has yet to be assessed but it seems likely that the transcriptional pattern alone will give a misleading impression of the role, if any, of this gene in the development of the nervous system.

Other additional sites of expression are observed with probes for these and other genes. Expression of *Kr* in the Malpighian tubules is consistent with early descriptions of the *Kr* phenotype and has recently been shown to be associated with a homoeotic transformation of this organ (Harbecke & Janning, 1989). Several other segmentation genes have been found to be expressed in the hindgut, though to date only *ftz* has been directly implicated in the differentiation of this tissue (Krause, Klemenz & Gehring, 1989).

Conclusions

The analysis of the patterns of expression of the *Drosophila* genes involved in embryonic pattern formation has provided many insights into their roles and interactions. The discovery and molecular characterisation of these genes has also provided a new impetus for the study of embryogenesis in other organisms, in particular vertebrates. A large number of cognates of the *Drosophila* segmentation genes have now been isolated from vertebrates and the expression patterns of many of these have been studied in some detail by *in situ* hybridisation (for a review see Dressler & Gruss, 1988). Whilst the validity of this approach is unquestionable, it is worth bearing in mind some of the limitations of such analyses as revealed by the *Drosophila* studies. Perhaps most important is the finding that many patterns of transcription are dynamic; in *Drosophila* there is no doubt that the knowledge of the mutant phenotypes of the genes contributed significantly to the interpretation of their expression patterns. Clearly there is a danger that in the absence of such information, false inferences might be drawn as to the time or place at which the expression of a gene is important. A particularly striking example of a spatially localised vertebrate gene product is provided by the *Vg-1* gene of *Xenopus*. In mature oocytes, the *Vg-1* transcript is restricted to the vegetal hemisphere and is tightly localised to the cortex. Following fertilisation, this localisation changes and the transcript becomes distributed in a graded fashion across the vegetal hemisphere. This distribution is highly suggestive of this gene product being required for an early determinative event such as mesoderm induction, an implication reinforced by the homology between the *Vg-1* protein and known mesoderm inducers (Weeks & Melton, 1987). Nevertheless, it is worth bearing in mind the gratuitous graded distributions of both *hb* and *cad* transcripts in the early *Drosophila* embryo when considering the significance of such transcripts. In the most extreme cases, as we have discussed, the spatially regulated expression of a gene may be without any functional significance. One way of pre-empting such fallacious inferences is to analyse the distribution of proteins rather than (or in addition to) transcripts. The most stringent test of the function of a gene, however, is, of course by its inactivation.

These rather negative considerations are offset by a very positive contribution of the *in situ* hybridisation technique to the analysis of *Drosophila* genes. As we have described, in a number of cases, previously unexpected multiple functions of several genes have been discovered by monitoring their expression throughout embryogenesis. This ability to discover additional requirements for a gene, normally masked by its mutant phenotype, has already had a significant impact on the study of neurogenesis in flies, and will likewise be of importance in the study of mutations in other organisms.

References

Akam, M. E. (1983). The location of *Ultrabithorax* transcripts in *Drosophila* tissue sections. *EMBO Journal*, **2**, 2075–84.

Akam, M. E. (1987). The molecular basis for pattern formation in the *Drosophila* embryo. *Development*, **101**, 1–22.

Baker, N. E. (1988). Localisation of transcripts from the *wingless* gene in whole *Drosophila* embryos. *Development*, **103**, 289–98.

Berleth, T., Burri, M., Thoma, G., Bopp, D., Richstein, S., Frigerio, G., Noll, M. & Nusslein-Volhard, C. (1988). The role of localisation of *bicoid* RNA in organising the anterior pattern of the *Drosophila* embryo. *The EMBO Journal*, **7**, 1749–56.

Carroll, S. B. & Scott, M. P. (1985). Localisation of the *fushi tarazu* protein during *Drosophila* embryogenesis. *Cell*, **43**, 47–57.

Carroll, S. B. & Scott, M. P. (1986). Zygotically active genes that affect the spatial expression of the *fushi tarazu* segmentation gene during early *Drosophila* embryogenesis. *Cell*, **45**, 113–26.

Cox, K. H., DeLeon, D. V., Angerer, L. M. & Angerer, R. C. (1984). Detection of mRNAs in sea urchin embryos by *in situ* hybridisation using assymetric RNA probes. *Developmental Biology*, **101**, 485–502.

Doe, C. Q. & Scott, M. P. (1988), Segmentation and homoeotic gene function in the developing nervous system of *Drosophila*. *Trends in Neurosciences*, **11**, 101–6.

Dressler, G. R. & Gruss, P. (1988). Do multigene families regulate vertebrate development? *Trends in Genetics*, **4**, 214–19.

Frasch, M. & Levine, M. (1987). Complementary patterns of *even-skipped* and *fushi tarazu* expression involve their differential regulation by a common set of segmentation genes in *Drosophila*. *Genes & Development*, **1**, 981–95.

Fronhofer, H.G. & Nusslein-Volhard, C. (1986). Organisation of anterior pattern in the *Drosophila* embryo by the maternal gene *bicoid*. *Nature (London)*, **324**, 120–5.

Garcia-Bellido, A. (1977). Homoeotic and atavic mutations in insects. *American Zoologist*, **17**, 613–29.

Gaul, U., Seifert, E., Schuh, R. & Jackle, H. (1987). Analysis of *Kruppel* protein distribution during early *Drosophila* development reveals posttranscriptional regulation, *Cell*, **50**, 639–47.

Hafen, E., Levine, M., Garber, R. L. & Gehring, W. J. (1983). An improved *in situ* hybridisation method for the detection of cellular RNAs in *Drosophila* tissue sections and its application for localizing transcripts of the homoeotic *Antennapedia* gene complex. *The EMBO Journal*, **2**, 617–23.

Harbecke, R. & Janning, W. (1989). The segmentation gene *Kruppel* of *Drosophila melanogaster* has homoeotic properties. *Genes & Development*, **3**, 114–22.

Howard, K. R. & Ingham, P. W. (1986) Regulatory interactions between the segmentation of genes *fushi tarazu*, *hairy*, and *engrailed* in the *Drosophila* blastoderm. *Cell*, **44**, 949–57.

Ingham, P. W. (1988). The molecular genetics of embryonic pattern formation in *Drosophila*. *Nature (London)*, **335**, 25–33.

Ingham, P. W., Howard, K. R. & Ish-Horowicz, D. (1985). Transcription patterns of the *Drosophila* segmentation gene *hairy*, *Nature (London)*, **318**, 439–45.

Ingham, P. W., Ish-Horowicz, D. & Howard, K. R. (1986) Correlative changes in homoeotic and segmentation gene expression in *Kruppel* mutant embryos of *Drosophila*. *The EMBO Journal*, **5**, 1659–65.

Ingham, P.W. & Martinez-Arias, A. (1986). The correct activation of Antennapedia and bithorax complex genes requires the *fushi tarazu* gene. *Nature (London)*, **324**, 592–7.

Jackle, H., Tautz, D., Schuh, R., Seifert, E. & Lehmann, R. (1986). Cross regulatory interactions among the gap genes of *Drosophila*. *Nature (London)*, **324**, 668–70.

Jurgens, G., Wieschaus, E., Nusslein-Volhard, C. & Kluding, H. (1984), Mutations affecting the pattern of the larval cuticle in *Drosophila melanogaster*. Zygotic loci on the third chromosome. *Roux's Archives*, **193**, 283–95.

Kilcherr, F., Baumgartner, S., Bopp, D., Frei, E. & Noll, M. (1986). Isolation of the *paired* gene of *Drosophila* and its spatial expression during early embryogenesis. *Nature (London)*, **321**, 493–9.

Knipple, D. C., Seifert, E., Rosenberg, U. B., Preiss, A. & Jackle, H. (1985). Spatial and temporal patterns of *Kruppel* gene expression in early *Drosophila* embryos. *Nature (London)*, **317**, 40–4.

Krause, H. M., Klemenz, R. & Gehring, W. (1988), Expression, modification and localisation of the *fushi tarazu* protein in *Drosophila* embryos. *Genes & Development*, **2**, 1021–36.

Kornberg, T., Siden, I., O'Farrell, P. & Simon, M. (1985). The *engrailed* locus of Drosophila: In situ localization of transcripts reveals compartment specific expression. *Cell*, **40**, 45–53.

Lehmann, R. & Nusslein-Volhard, C. (1987). *Hunchback*, a gene required for segmentation of an anterior and posterior region of the *Drosophila* embryo. *Developmental Biology*, **119**, 402–17.

Lemotte, P. K., Kuroiwa, A., Fessler, L. I. & Gehring, W. J. (1989). The homoeotic gene *Sex Combs Reduced* of *Drosophila*: gene structure and embryonic expression. *The EMBO Journal*, **8**, 219–27.

Levine, M., Hafen, E., Garber, L. & Gehring, W. J. (1983). Spatial distribution of *Antennapedia* transcripts during *Drosophila* development. *The EMBO Journal*, **2**, 2037–46.

Lewis, E. B. (1963). Genes and developmental pathways. *American Zoologist*, **3**, 33–56.

Lewis, E. B. (1978). A gene complex controlling segmentation in *Drosophila*. *Nature (London)*, **276**, 565–70.

Macdonald, P., Ingham, P. & Struhl, G. (1986). Isolation. Structure and expression of *even-skipped*: a second pair-rule gene of *Drosophila* containing a homeo box. *Cell*, **47**, 721–34.

Macdonald, P. & Struhl, G. (1986). A molecular gradient in early *Drosophila* embryos and its role in specifying the body pattern. *Nature (London)*, **324**, 537–54.

Mahoney, P. A. & Lengyel, J. A. (1987). The zygotic segmentation mutant *tailless* alters the blastoderm fate map of the *Drosophila* embryo. *Developmental Biology*, **122**, 464–70.

Martinez-Arias. A., Ingham, P. W., Scott, M. P. & Akam, M. E. (1987). The spatial and temporal deployment of *Dfd* and *Scr* transcripts throughout development of *Drosophila*. *Development*, **100**, 673–83.

Mlodzig, M., Fjose, A. & Gehring, W. J. (1985). Isolation of *caudal*, a *Drosophila* homeo box-containing gene with maternal expression, whose transcripts form a concentration gradient at the pre-blastoderm stage. *The EMBO Journal*, **4**, 2961–9.

Nusslein-Volhard, C. & Wieschaus, E. (1980). Mutations affecting segment number and polarity in *Drosophila, Nature (London)*, **287**, 795–801.

Nusslein-Volhard, C., Wieschaus, E., & Kluding, H. (1984). Mutations affecting the pattern of the larval cuticle in *Drosophila melanogaster*. I. Zygotic loci on the second chromosome. *Roux's Archives*, **193**, 267–82.

Sander, K. (1976). Specification of the basic body pattern in insect embryogenensis. *Advances in Insect Physiology*, **12**, 125–38.

Struhl, G. (1983). Role of the esc^+ gene product in ensuring the selective repression of segment-specific homeotic genes in *Drosophila*. *Journal of Embryology and Experimental Morphology*, **76**, 297–331.

Tautz, D., Lehmann, R., Schnurch, H., Schu, R., Seifert, E., Kienlin, A., Jones, K. & Jackle, H. (1987). Finger protein of novel structure encoded by *hunchback*, a second member of the gap class of *Drosophila* segmentation genes. *Nature (London)*, **327**, 383–9.

Weeks, D. L. & Melton, D. A. (1987). A maternal mRNA localised to the vegetal hemisphere in Xenopus eggs codes for a growth factor related to TGF-β, *Cell*, **51**, 861–7.

Wieschaus, E., Nusslein-Volhard, C. & Jurgens, G. (1984). Mutations affecting the pattern of the larval cuticle in *Drosophila melanogaster* III. Zygotic loci on the X-chromosome and fourth chromosome. *Roux's Archives*, **193**, 296–307.

H. Perry-O'Keefe, C. R. Kintner, J. Yisraeli, and D. A. Melton

The use of *in situ* hybridisation to study the localisation of maternal mRNAs during *Xenopus* oogenesis

Introduction

A method for *in situ* hybridisation using *Xenopus laevis* oocytes is described in detail. Using this technique, we have studied the distribution of several localised maternal mRNAs during *Xenopus* oogenesis. Our results suggest that all of the RNAs tested, including mRNAs localised to either the animal or vegetal ends of oocytes, are evenly distributed during early oogenesis. Localisation begins during the middle stages of oogenesis, and full-grown oocytes demonstrate the most striking localisation.

One parameter that can be used to identify a gene whose product is developmentally important is the spatial pattern of its transcript. *In situ* hybridisation is a method commonly used for determining the spatial localisation of any relatively abundant gene transcript. Although this procedure cannot be used to detect low abundance mRNAs, and can be hindered by fixation artefacts, it can be a very useful method, particularly in conjunction with other techniques.

There are many experiments in which *in situ* hybridisation has been used to study how genes are expressed during development. For example, these experiments have been particularly informative with respect to the expression of homeobox genes of *Drosophila* (for review see Akam, 1987). In these cases this technique has been utilised in conjunction with the analysis of mutant phenotypes. *In situ* hybridisation has also been used in sea urchins with histone genes (Angerer *et al.*, 1984; Venezky *et al.*, 1981; Cox *et al.*, 1984) transcripts of which are localised to the pronucleus until its breakdown following meiotic prophase. In addition, this technique has been instrumental in determining the localisation of the Spec 1 transcripts which are found in dorsal ectodermal cells of the sea urchin pluteus larvae (Lynn *et al.*, 1983). Finally, in the case of *Xenopus*, *in situ* hybridisation has been used to determine the location of the histone genes in oocytes (Jamrich *et al.*, 1984), the localisation of neural cell-adhesion molecule (Kintner & Melton, 1987), as well as the developmental timing and the general pattern of the localisation of the *Vg1* gene (Melton, 1987).

The unusually large cell size of the *Xenopus* egg and early embryonic blastomeres, in addition to a yolk rich cytoplasm, pose a special problem for *in situ* hybridisation. A method which is satisfactory, but not ideal, has been described (Melton, 1987) and its rationale and application is more fully described here. In addition, we describe results from ongoing studies of the spatial localisation of the *Xenopus* localised clones *An1*, *An2* and *An3*. Previous studies have demonstrated that the *An1*, *An2*, and *An3* clones are localised to the animal pole of the oocyte. This was shown by cutting the top and bottom quarters from oocytes, isolating RNA for Northern blots from the respective poles, and probing those blots with the three different animal-specific clones (Rebagliati *et al.*, 1985; Weeks *et al.*, 1985). In this study, the histone *H4* and *Vg1* clone were used as standards, the former to show an mRNA with even distribution, and the latter to demonstrate a localised mRNA. As expected, the *Vg1*, *An1*, *An2* and *An3* clones exhibited localisation during oogenesis, and the histone mRNA was evenly distributed. One of the most interesting characteristics noted is that all the maternal mRNAs studied show an even distribution in early oogenesis. As oogenesis proceeds some mRNAs gradually become localised. This suggests that these mRNAs may all be localised using the same cellular mechanism.

Results

Early oogenesis – stages I and II

Small pre-vitellogenic oocytes (stages I and II, Dumont, 1972) were collected, fixed, sectioned and used for *in situ* hybridisation with a variety of RNA probes (Fig. 1). In all cases, hybridisation with histone H4 probe was used as a control. As shown previously (Jamrich *et al.*, 1984), histone mRNA is evenly distributed in the cytosol at all stages of oogenesis. The present results show that each RNA, (*Vg1*, *An1*, *An2*, and *An3*), is evenly distributed in small oocytes (Fig. 2). It is interesting to note that all of the probes appear evenly distributed throughout the cytoplasm, despite the formal possibility that some of these sequences might be localised as they were being synthesised and stored in the small oocytes. Rather it appears that all of the sequences that we are studying are first synthesised and uniformly distributed in young oocytes and then localised at a later point in development. All of the probes used give a very strong signal primarily because the mRNA is concentrated in these small, largely yolk-free, cells.

Mid-oogenesis – stages III and IV

In situ hybridisations were done using middle stage (III–IV) oocytes (Dumont, 1972) with the following probes: *Vg1*, *An1*, *An2*, and *An3* (see Fig. 1). The results demonstrate that by the middle stages of oogenesis

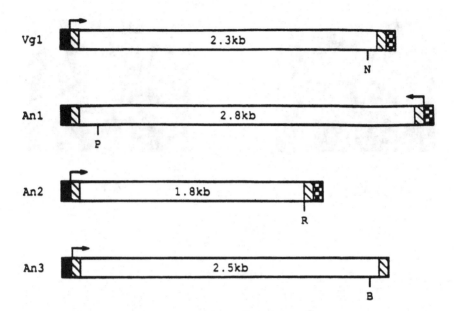

Figure 1. Probes used for *in situ* hybridisations in *Xenopus* oocytes. Schematic diagrams of the probes that we used for *in situ* hybridisation of *Xenopus* oocytes. In each case the black box represents the promoter for *SP6* RNA polymerase, the checkered boxes represents the promoter for *T7* RNA polymerase, and the striped boxes represent the polylinker region. The size of each transcript is indicated. All clones were subcloned into to the pSP70 vector (Krieg & Melton, 1987) with the exception of *An3* which is in pSP65 (Melton *et al.*, 1984). The arrow denotes the direction of transcript in order to give an antisense RNA probe, and the letters denote the restriction enzymes which were used to linearise the transcripts: B, Bam HI; N, Nco I; P, Pvu II; and R, Eco RV.

localisation of some mRNAs has already begun (Fig. 3). The H4 RNA remains evenly distributed, although the signal seems to decrease somewhat (not shown). This is true for all of the RNAs and is due to the increased size of the oocyte, which (given the same amount of RNA) 'dilutes' the target. The other standard, *Vg1*, is already beginning to be localised. The signal is found both surrounding the germinal vesicle as well as beginning to be localised to one pole of the oocyte. Similarly, *An1*, *An2*, and *An3* mRNAs have also begun to be localised (Fig. 3), all three beginning to be more concentrated at one end of the oocyte (presumably the animal pole). Although mid-oogenesis spans a number of days, it is interesting to note that *An2* and *An3* appear to follow the same pattern of deposition as *Vg1* (Fig. 3). The sections probed

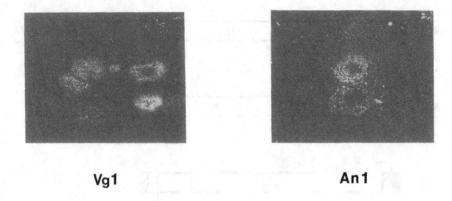

Vg1 An1

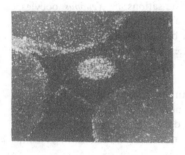

An2 An3

Figure 2. *In situ* hybridisation with pre-vitellogenic *Xenopus* oocytes. *In situ* hybridisations were done using a variety of ^{32}P labelled single stranded RNA probes and tissue sections from small, (stage I and II – Dumont, 1972), albino oocytes. The probes that were used are indicated under each photograph. The germinal vesicles are seen as a black hole in the centre of some of the oocytes. All photographs were taken in dark field. Note that all probes give an evenly distributed signal.

with the animal specific clones demonstrate a signal which is widely distributed throughout the cytoplasm although clearly skewed toward one pole.

Late oogenesis – stages V and VI/work in progress

Late stage oocytes (V–VI) were used for *in situ* hybridisations with the same probes (see Fig. 1). In general, the results of *in situ* hybridisations done on late stage oocytes were similar, but more pronounced, to those seen for the middle stage oocytes (Fig. 4). The concentration of the *Vg1* mRNA is much greater in that hemisphere of the oocyte (Fig. 4). One can clearly see that the signal is largely confined to a tight crescent in the vegetal hemisphere of the oocyte. In contrast, *An2* and *An3* both show a gradient of signal strongest in the animal hemisphere (Fig. 4). Although the abundance of the mRNAs clearly varies, *Vg1*, *An2* and *An3* all display a particular localisation with the stage VI oocyte.

Another notable difference between the *An2*, *An3*, and *Vg1*, is that the *Vg1* mRNA is located at or near the cortex of the oocyte. Neither *An2* nor *An3* give any indication of a cortical localisation. For example, the higher magnification photographs in Figure 4(d) and (e) demonstrate that *An3* mRNA is not localised to the cortex of the oocyte, but rather is distributed in a gradient along the animal–vegetal axis.

Discussion

We have characterised the localisation of four mRNAs in pre-vitellogenic oocytes, demonstrating that they are all apparently evenly distributed throughout the cytoplasm. One of our long-term objectives is to address the question of how a cell localises specific mRNAs.

In situ hybridisation has been established as a reliable method for determining mRNA localisation in a variety of organisms, including flies, sea urchins, mice, and frogs. In this study, a series of cDNAs have been used to study the localisation of mRNAs during frog oogenesis. The *H4* and *Vg1* cDNAs were chosen as controls, one (*H4*) being evenly distributed throughout oogenesis and the other (*Vg1*) becoming localised during the middle and late stages of oogenesis (Jamrich *et al.*, 1984; Rebagliati *et al.*, 1985; Melton, 1987).

The results reported here suggest that the *An1*, *An2*, and *An3* mRNAs become localised to the oocyte animal hemisphere. This localisation process apparently begins during the middle stages of oogenesis and continues throughout the late stages. It would be interesting to know if this localisation persists through subsequent development, i.e. after fertilisation.

The question of the mechanism of localisation is indeed a complicated one. Although these experiments do not directly address either the precise timing

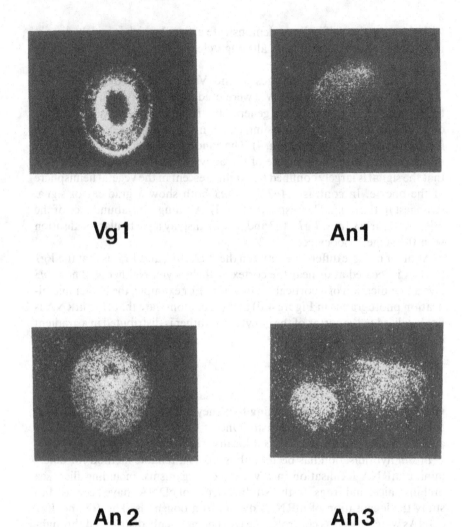

Figure 3. *In situ* hybridisations of mid-stage *Xenopus* oocytes. Middle stage albino oocytes (stages III and IV, Dumont, 1972) were probed with a variety of antisense RNAs, as indicated under each photograph. All probes were labelled to a high specific activity with ^{32}P. Photographs were taken in dark field.

Note that the *H4* probe is evenly distributed, while the *Vg1* is being concentrated to one end of the oocyte (in this case the bottom two-thirds of the oocyte). Likewise the *An1*, *An2* and *An3* probes indicate that the corresponding mRNAs are beginning to be concentrated at one end of the oocyte (in these cases the top of the oocyte).

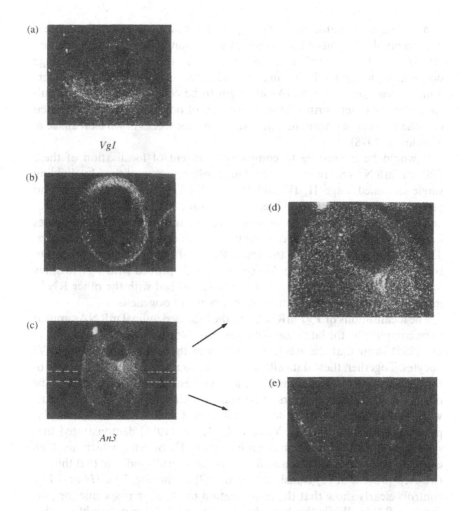

Figure 4. *In situ* hybridisation using fully grown *Xenopus* oocytes. Late stage albino oocytes (stages V and VI, Dumont, 1972), were used for *in situ* hybridisation. All probes (antisense RNA transcripts) were labelled to a high specific activity with ^{32}P. The probes that were used are indicated under each photograph. All photographs were taken in dark field. The germinal vesicle appears as a black hole in the oocyte. (a) *Vg1* probe indicates that the *Vg1* mRNA is concentrating in a tight crescent around one hemisphere of the oocyte. *An2* (b) and *An3* (c) probes show a concentration of the corresponding mRNAs in the animal hemisphere. This is more obvious in the enlargements of the photo in (c). The oocyte shown in (c) was cut to give the enlargements shown in panels (d), (top) and (e) (bottom).

or mechanism of localisation they do give an indication that a similar process may be involved in all of the cases that were examined. All of the localised mRNAs, *Vg1*, *An1*, *An2*, and *An3*, are evenly distributed in early stage oocytes. Although the beginning of localisation was not specifically determined, these localised mRNAs all begin to be concentrated in one hemisphere or the other during the middle stages of oogenesis. It must be noted that the process of oogenesis can take up to four weeks *in vitro* (Wallace & Misulovin, 1978).

It would be interesting to compare the extent of localisation of these different mRNAs more accurately. This could be accomplished by dividing single-sectioned stage II, III and IV oocytes between several slides and probing each slide with a different RNA probe. Although this would not determine the definite time at which each individual mRNA becomes localised, it would help to determine the relative order in which this occurs. That is, if the *Vg1* mRNA is the first mRNA (of the five examined here) to begin to be localised, one would expect the slide probed with *Vg1* to show localisation of that message, while the slides probed with the other RNAs might not show any localisation at that point in oogenesis.

The localisations of *Vg1* mRNA and the localised animal mRNAs appear to be complete by the late stages of oogenesis. Northern blots (Rebagliati *et al.*, 1985) show that the RNA is localised in the animal pole in stage VI oocytes. Together, these data allow one to rule out the possibility that any of the localised mRNAs examined here are synthesised in only one part of the oocyte. Rather, there appears to be some cellular mechanism which must work to actively concentrate some evenly distributed mRNAs to particular parts of the oocyte. Indeed, Yisraeli & Melton (1988) demonstrated that synthetic *Vg1* and mRNA injected into stage III oocytes, which are then cultured, will be properly localised. Preliminary results indicate that this is a two-step process of (1) translocation and (2) anchoring. The *H4* and *Vg1* controls clearly show that the translocation mechanism is specific for particular mRNAs. While this hypothesis is favoured, other possibilities exist. For example, in principle, localised degradation of the RNA could explain the observed localisation. However, we have no results that indicate that different methods of localisation are utilised for any of these mRNAs. To the contrary, the localisation looks strikingly similar for the animal specific clones. It will be interesting to learn if the synthetic animal specific mRNAs can be localised when injected into small oocytes which are then cultured *in vitro*. If the mechanism of localisation does occur by the same method for all four localised RNAs it would be intriguing to determine and compare the different localisation signals.

Materials and methods

The method detailed below is a combination of several published methods (Brahic & Hasse, 1978; Hafen et al., 1984; Jamrich et al., 1984; Dworkin-Rastl et al., 1986; Jamrich et al., 1987) and is extensively derived from the work of the Angerers and their colleagues (Angerer & Angerer, 1981; Lynn et al., 1983; Angerer et al., 1984; and Cox et al., 1984). In several instances, steps have been included, excluded or substantially modified to meet the special requirements of Xenopus oocytes. The rationale for these changes is discussed below.

Animals

Albino *Xenopus laevis* frogs were purchased from Xenopus I, (Ann Arbor, MI). Oocytes were obtained by anesthetising female frogs and removing part of the ovarian tissue. The individual oocytes were then separated from one another and sized according to Dumont (1972).

All oocytes used for *in situ* hybridisations were albino in order to avoid confusion of silver grains and pigment granules which look similar under both light and dark field microscopy. Orientation with respect to the animal–vegetal axis depended on the concentration of yolk platelets (highest in the vegetal hemisphere of stage IV–VI oocytes), as well as the position of the germinal vesicle which becomes displaced to the animal hemisphere by stage IV.

Reagents

A paraformaldehyde stock (5%) is made by dissolving paraformaldehyde at 60°C in water and adjusting the pH to 8 with NaOH. The solution is filtered through Whatman paper and stored at 4°C. A 20X PBS stock is made of 2.6 M NaCl, 140 mM Na_2HPO_4, 60 mM NaH_2PO_4. A 20X SSPE stock is made of 3.6 M NaCl, 200 mM NaH_2PO_4, pH 7.4, 20 mM EDTA. A 2.5 M triethanolamine stock is adjusted to pH 8.0.

Radionucleotides were purchased from Amersham and restriction enzymes and RNA polymerases were from Promega Biotec.

Fixation

A variety of fixatives were tested. The principal problem is to identify a fixative that retains the RNA and at the same time provides good tissue morphology. For most embryonic stages a number of fixatives give satisfactory results (Jamrich et al., 1984; Dworkin-Rastl et al., 1986). However, the oocytes and eggs of *Xenopus* pose a special problem. These cells are large and unusually yolky making it difficult to obtain good morphology and RNA retention.

To test a fixative for RNA retention, ^{32}P-labelled globin mRNA was injected into oocytes and the injected cells were fixed one hour later. The amount of radiolabel retained after removing the fixative and dehydration in ethanol was used as a measure of RNA retention. In addition, morphology was scored by inspection of stained sections. For those fixatives that passed this first test, RNA retention in sections was measured by hybridisation with an antisense histone RNA probe. This probe hybridises to an abundant endogenous mRNA coding for the histone H4.

Initially standard fixatives were tested including Bouin's, alcoholic Bouin's, Smith's, Peryeni's, and San Felice's all of which are described in Humason (1972). The paraformaldehyde–heptane fixative that works so well for *Drosophila* (Hafen *et al.*, 1984) and the gluteraldehyde fixative that is often used for sea urchins (Cox *et al.*, 1984) were also tried. Following the work of Gall and his colleagues (Jamrich *et al.*, 1984), formaldehyde and paraformaldehyde were tested at a variety of concentrations (1, 2, 3, 4, 5, and 10%) in 0.25 M NaCl, with and without glacial acetic acid (5%). Freeze substitution (Gurdon *et al.*, 1976) was also tried.

In those few cases where satisfactory results were obtained, further experiments were performed to test slight variations in the concentration of the components to find the optimal fixative. After numerous tests, we devised a fixative, a modification of those previously reported, that penetrates amphibian tissue quickly and fixes the RNA effectively. A mixture of ethanol:acetic acid:chromium trioxide (95:5:0.25; v:v:v) gives the best autoradiographic signal, while still retaining good tissue morphology. The advantage of this fixative is especially apparent when used for oocytes and eggs, but it is equally effective on later embryonic stages. We also find it helpful to re-fix the tissue briefly in paraformaldehyde (Hafen *et al.*, 1984) following the removal of the paraplast from sectioned specimens that are stuck to slides.

Fixative is prepared by cooling 95% ethanol, 5% glacial acetic acid on ice and adding 50% (w:v) chromium trioxide to a final concentration of 0.25% immediately before use. Oocytes or embryos are added to ice cold fixative and mixed occasionally over a period of one hour. The solution is kept at 0–4°C during fixation. The fixative is removed with an increasing series of ice cold ethanol rinses and the specimens are embedded in paraplast. This regimen is somewhat flexible:

> 95% ethanol, two to three changes for a total of 20–30 min, 0–4°C
> 100% ethanol, two changes for a total of 20–30 min, 0–4°C
> 50% ethanol, 50% xylene, 10–15 min, room temperature
> 50% xylene, 50% paraplast (VWR), 1 h, 50°C
> Paraplast, two changes, for a total of at least 4 h (usually overnight), 50°C
> Embedded in paraplast

Sectioning

It is often helpful to soak yolk-rich tissues in glycerol so that they become less brittle for sectioning. This can be accomplished by cutting sections in the paraplast until the knife reaches the tissue, and then continuing for a short distance (~ 30 μm). The block is then soaked overnight in 5% glycerol.

Mounting sections

We found that the most commonly used methods for mounting sections on slides resulted in a loss of the sections during the procedures leading up to hybridisation. We tried both subbing slides (Gall & Pardue, 1971) and polylysine treatment of slides. Following a suggestion from Dr Rolf Zeller we use silane treated slides (Gottlieb & Glaser, 1975). Slides are first washed in hot, soapy water, rinsed thoroughly in hot, tap water, followed by a final rinse in 95% ethanol, 0.1% acetic acid. After drying, the slides are soaked for 1 h in 10% silane (3-aminopropyl triethoxysilane, Sigma) at 80–90°C in a vented oven. Slides are rinsed in toluene, in a hood, and air dried. The dry slides are incubated in 4% paraformaldehyde for 1 h at room temperature and wiped clean with a tissue immediately before use.

6–8 μ sections are cut on a rotary microtome with metal knives and ribbons are floated onto DEPC-treated water on the silane treated slides. After the ribbons spread, excess water is removed and the slides are allowed to dry on a warm plate (40°C) for 12–72 h.

Pre-hybridisation treatments

Several pre-hybridisation treatments have been reported to enhance the autoradiographic signal:noise ratio. These include treatment with proteinase K to remove some protein from the fixed sections (Angerer & Angerer, 1981), treatment with 0.2M HCl to remove basic proteins, and finally with acetic anhydride (Hayashi *et al.*, 1978) to neutralise positive charges on the section and the slide. While all of these are routinely included in our protocol, only titration with varying amounts of proteinase K has been systematically tested. The other treatments have been included without testing for their effectiveness.

The length of time for digestion with proteinase K varies with the amount of tissue (perhaps yolk) present. We find that small oocytes (stages I–III) should be treated for 15 min, whereas large oocytes and eggs should be treated for 30–40 min. In general, embryos and tadpoles are treated for 30 min.

Paraplast is removed by soaking slides in two changes of xylene, 10 min each. Tissue sections are rehydrated quickly by dipping in 100%, 95%, 80%,

70%, and 40% ethanol for 30–60 s each. The slides are immediately placed in 4% paraformaldehyde, 1X PBs for 20 min.

Following the refixation with paraformaldehyde, slides are rinsed in 2 × SSPE and transferred to a proteinase K solution for 15–30 min at 37°C. The proteinase K is used at 3 μg/ml in 0.1 M Tris, pH 7.5, 0.01 M EDTA. Slides are rinsed in 2 × SSPE and then incubated in 0.2m HCl for 15 min at room temperature.

Slides are again rinsed in 2 × SSPE and then treated with acetic anhydride (Hayashi et al., 1978). The slides are placed in 0.1 M triethanolamine, pH 8.0 and acetic anhydride is added to a concentration of 0.25% while the slides are agitated in the solution. Five minutes later the same amount of acetic anhydride is again added and the slides are again agitated.

Slides are rinsed in 2 × SSPE and held (up to several hours) in this solution until the probe is ready for addition.

Probe synthesis

The synthesis of single stranded RNAs for use as hybridisation probes has been described previously (Melton et al., 1984). The advantages of these RNA probes over nick-translated DNAs for in situ hybridisation was first shown and fully documented by Cox et al. (1984). Consistent with their results, we find that single-stranded RNA probes are preferable to nick-translated probes both in terms of sensitivity and signal:noise ratio.

Radiolabelled probes are prepared by synthesis with SP6 or T7 RNA polymerase using a $^{32}P\alpha UTP$ (400 Ci/mmol). Typically transcription reactions are set up in 10 μl with 50 μCi of radiolabelled UTP. 80–90% of the added label is incorporated into RNA transcripts. The specific activity of this RNA is about 100 C_i/mmol or about 6×10^8 dpm/μg. Further details on the transcription reaction can be found in Krieg & Melton (1987).

The RNA probes are hydrolysed in 40 mM $NaHCO_3$, 60 mM Na_2CO_3 at 60°C top reduce the size of the RNA to an average length of about 150 nucleotides (Lynn et al., 1983). For RNAs about 0.5 kb long, the incubation time is 42 minutes, for RNAs 1 kb long, 50 minutes, and RBAs 2 kb long require 56 minutes. Following hydrolysis and neutralisation, typically with a final concentration of 5% glacial acetic acid and 0.3 M sodium acetate pH 4.8, the RNA is recovered by ethanol precipitation with 10 μg carrier RNA.

Hybridisation conditions

In order to saturate target RNAs, the radiolabelled RNA probes should be added to the hybridisation solution at a final concentration of about 0.2–0.3 μg/ml per kb probe complexity (Cox et al., 1984). In practice, we add about 3×10^7 cts/min of probe to 50–200 μl of hybridisation solution. The amount of probe used for each slide varied depending on the probe. The primary factor considered was the abundance of the mRNA. If it is a low abundance mRNA, the amount of labelled probe may be decreased by as much as 10-fold. Depending on the amount of tissue on a slide, 10–100 µl of hybridisation solution is added to each slide. The final concentration of components in the hybridisation solution is 50% deionised formamide, 0.3 M NaCl, 10% dextran sulphate, 5 mM Tris-HCl, Ph 8.0, 0.2 mM EDTA, 1 × Denhardts (0.02% each of BSA, Ficoll, polyvinylpyrolidine), carrier RNA (100–500 µg/ml), and radiolabelled probe. When using ^{35}S-labelled probes, the hybridisation solution should contain 100 mM β-mercaptoethanol or 10 mM DTT to reduce background. We prepare the hybridisation solution fresh before each application, with all of the components except the radiolabelled probe and carrier tRNA. The latter solution is heated to 80°C to denature the RNA and then added to the hybridisation solution just before application to the tissue sections.

Slides are dipped in DEPC-treated water and allowed to dry until most of the water is gone but the tissue is still moist. It is critical that the tissue is *not* dry, otherwise the probe will not penetrate the tissue well. Following the probe addition, the sections are covered with a coverslip and placed in mineral oil at 50°C. Hybridisation proceeds for 10–15 h at 50°C.

Washing conditions

The stringency of the wash conditions that can be tolerated depends on the probes being used and their lengths of homology with target sequences. In general, a stringent wash in 0.1 × SSPE at 55°C for 1 h is quite effective at reducing backgrounds without loss of signal.

Mineral oil is removed by three rinses (5 minutes each) in chloroform. The coverslips are allowed to fall off while the slides sit in 4 × SSPE. For ^{35}S-labelled probes, it is helpful to include 100 mM β-mercaptoethanol or 10 mM DTT when removing the coverslips and washing the slides, in order to reduce non-specific background.

Slides are washed for 1 h at room temperature in 2 × SSPE.

Slides are treated with RNase A (Sigma) to remove unhybridised probe. Incubate slides in 4 × SSPE, RNase A (20 µg/ml), for 30–45 min at 37°C.

The slides are then subjected to a more stringent wash in 0.1 × SSPE for 1 h at 55–60°C This is normally done in 1 l of circulating solution. For ^{35}S-labelled probes the stringent wash is performed in 50% formamide, 2 × SSPE at 50°C for 1 h.

Autoradiography

Glycerol is included for dipping in emulsion because it greatly reduces autoradiographic artefacts that come from emulsion stretching and cracking. Some protocols suggest drying slides after hybridisation, before dipping into emulsion. On the contrary, we find that it is important to keep the slide wet at all times in order to reduce the probability of trapping air bubbles in the tissue, particularly the yolk. Thus, the tissue is never allowed to dry from the time it is rehydrated, after removing the paraplast until it is dipped in emulsion.

Washed slides are kept at room temperature in 0.3 M ammonium acetate, 1% glycerol before dipping into emulsion. Kodak NTB2 emulsion is prepared by incubating 5 ml of the stock (stored at 4°C) at 42°C, and then diluting it 1:1 (v:v) with 0.6 M ammonium acetate, 2% glycerol at 42°C. The mixture is placed in a dipping chamber and held at 42°C. Emulsion-coated slides are kept in a large light proof box and allowed to dry for 1 hour. Slides are then transferred to a slide box containing desiccant. The latter box is sealed carefully and kept at 4°C for the duration of the exposure (5–12 days).

Slides are developed at 15–18°C in Kodak D19 developer, diluted 1:1 with H_2O, for 2.5 min, rinsed for 10 s in 0.2% acetic acid, and fixed in Kodak fixer for 5 min. Slides are then washed for at least 5 min in cold, tap water and prepared for staining.

Staining and mounting

A variety of stains can be used though one should be careful about using stains which require an acid treatment because this can cause the emulsion to fog. For most purposes a simple and effective stain is Giemsa (Polysciences). Giemsa stain is diluted 1:30 (v:v) in 50 mM PO_4 buffer pH 6.5, and slides are submerged in stain for 10 min. Following washing with running cold, tap water, the slides are destained in 70% ethanol and dehydrated through an ethanol series, and finally into xylene. Coverslips are placed on slides still wet with xylene. The mounting media is a mixture of Canada balsam (BDH Chemicals) and methylsalicylate.

References

Akam, M. (1987). The molecular basis for metameric pattern in the Drosophila embryo. *Development*, **101**, 1–22.

Angerer, L. M. & Angerer, R. C. (1981). Detection of PolyA$^+$ RNA in sea urchin eggs and embryos by quantitative in situ hybridization. *Nucleic Acids Research*, **9**, 2819–40.

Angerer, L. M., DeLeon, D. V., Angerer, R. C., Showman, R. M., Wells, D. E. & Raff, R. A. (1984). Delayed accumulation of maternal histone mRNA during sea urchin oogenesis. *Developmental Biology*, **101**, 477–84.

Brahic, M. & Haase, A. T. (1978). Detection of viral sequences of low reiteration frequency by *in situ* hybridization. *Proceedings of the National Academy of Sciences, USA*, **75**, 6125–9.

Cox, K. H., DeLeon, D. V., Angerer, L. M. & Angerer, R. C. (1984). Detection of mRNAs in sea urchin embryos by *in situ* hybridization using asymmetric RNA probes. *Developmental Biology*, **101**, 485–502.

Dumont, J. N. (1972). Oogenesis in *Xenopus laevis* (Daudin). *Journal of Morphology*, **136**, 153-80.
Dworkin-Rastl, E., Kelley, D. B. & Dworkin, M.B. (1986). Localization of specific mRNA sequences in *Xenopus laevis* embryos by *in situ* hybridization. *Journal of Embryology and Experimental Morphology*, **91**, 153-68.
Gall, J. G. & Pardue, M. L. (1971). Nucleic acid hybridization in cytological preparations. In *Methods in Enzymology* (eds L. Grossman & K. Moldave) vol. 21, pp. 470-80. Academic Press.
Gottlieb, D. I. & Glaser, L. (1975). A novel assay of neuronal cell adhesion. *Biochemical Biophysical Research Communications*, **63**, 815-21.
Gurdon, J. B., Partington, G. A. & De Robertis, E. M. (1976). Injected nuclei in frog oocytes: RNA synthesis and protein exchange. *Journal of Embryology and Experimental Morphology*, **36**, 541-53.
Hafen, E., Kuroiwa, A. & Gehring, W. J. (1984). Spatial distribution of transcripts from the segmentation gene *fushi tarazu* during *Drosophila* embryonic development. *Cell* **37**, 833-41.
Hayashi, S., Gillam, I. C., Delaney, A. D. & Tener, G. M. (1978). Acetylation of chromosome squashes of *Drosophila melanogaster* decreases the background in autoradiographs from hybridization with ^{125}I labelled RNA. *Journal of Histochemistry Cytochemistry*, **26**, 677-9.
Humason, G. L. (1971). *Animal Tissue Techniques*. W. H. Freeman and Co. (San Francisco).
Jamrich, M., Mahon, K. A., Gavis, E. R. & Gall, J. G. (1984). Histone RNA in amphibian oocytes visualized by *in situ* hybridization to methacrylate-embedded tissue sections. *The EMBO Journal*, **3**, 1939-43.
Jamrich, M., Sargent, T. D. & Dawid, I. B. (1987). Cell-type-specific cytokeratin genes during gastrulation of *Xenopus laevis*. *Genes in Development*, **1**, 124-32.
Kintner, C. R. & Melton, D. A. (1987). Expression of *Xenopus* N-CAM RNA in ectoderm is an early response to neural induction. *Development*, **99**, 311-25.
Krieg, P. A. & Melton, D. A. (1987). *In vitro* RNA synthesis with SP6 RNA polymerase. In *Methods in Enzymology* (eds J. N. Abelson and M. I, Simon), vol. 155 pp. 397-415, New York; Academic Press.
Lynn, D. A., Angerer, L. M., Bruskin, A. M., Klein, W. H. & Angerer, R. C. (1983). Localization of a family of mRNAs in a single cell type and its precursors in sea urchin embryos. *Proceedings of the National Academy of Sciences, USA*, **80**, 2656-60.
Melton, D. A. (1987). Translocation of a localized maternal mRNA to the vegetal pole of *Xenopus* oocytes. *Nature (London)*, **328**, 80-2.
Melton, D. A., Krieg, P. A., Rebagliati, M. R., Maniatis, T., Zinn, K. & Green, M. R. (1984). Efficient *in vitro* synthesis of biologically active RNA and RNA hybridization probes from plasmics containing a bacteriophage SP6 promoter. *Nucleic Acids Research*, **12**, 7035-56.
Rebagliati, M. R., Weeks, D. L., Harvey, R. P. & Melton. D. M. (1985). Identification and cloning of localized maternal RNAs from *Xenopus* eggs. *Cell*, **42**, 769-77.

Venezky, D. L., Angerer, L. M. & Angerer, R. C. (1981). Accumulation of histone repeat transcripts in the sea urchin egg pronucleus. *Cell*, **24**, 385–91.

Wallace, R. A. & Misulovin, Z. (1978). Long-term growth and differentiation of *Xenopus* oocytes in a defined medium. *Proceedings of The National Academy of Sciences, USA*, **75**, 5534–8.

Weeks, D. L. & Melton, D. M. (1987). A maternal mRNA localized to the vegetal hemisphere in *Xenopus* eggs codes for a growth factor related to TGF-β. *Cell*, **51**, 861–6.

Weeks, D. L., Rebagliati, M. R., Harvey, R. P. & Melton, D. A. (1985). Localized maternal nMRAs in *Xenopus laevis* eggs. *Cold Spring Harbor Symposia in Quantitative Biology*, **50**, 21–9.

Yisraeli, J. K. & Melton, D. M. (1988). The maternal mRNA *Vg1* is correctly localized following injection into *Xenopus* oocytes. *Nature (London)*, **336**, 592–5.

David G. Wilkinson

In situ hybridisation in the analysis of genes with potential roles in mouse embryogenesis

Introduction

The mouse is widely used as a system for the study of mammalian development, and many studies have documented the morphological events that occur as tissues first form and become organised in the post-implantation embryo. However, very little is known of the molecular mechanisms that underlie these morphogenetic processes. In large part this ignorance is due to technical problems in studying embryos that are small, relatively inaccessible, and in which systematic genetics is difficult. One approach towards understanding development is to identify and analyse the function of genes that may have roles in the early events of the formation and spatial organisation of tissues. There are three phases to this type of approach; first, to select candidate genes suspected of having potential roles in development; second, to study their expression pattern in order to gain clues as to what process (or processes) they may be involved in; and third, to analyse their function by manipulative experiments guided by knowledge of their normal expression patterns. Below, we will discuss some of the ways that candidate developmental genes have been selected, and their expression patterns studied by *in situ* hybridisation.

Selection of candidate developmental genes
Genetics

The most direct way to identify genes with developmental roles is through the use of genetics, an approach that has been conspicuously successful for the analysis of *Drosophila* embryogenesis (Akam, 1987; Ingham, 1988). A large number of genetic loci have been identified that affect various aspects of development in mice. Many of these loci may encode products with rather widespread physiological roles that disrupt development in a pleitropic manner, but at least some may be directly involved in patterning tissues. A drawback of this approach is the time-consuming effort required to clone the corresponding gene, although this is considerably aided by strategies using insertional mutagenesis.

Homologies with developmental genes from other systems

The most popular approach in recent years has been inspired by the finding that certain insect developmental genes, found by the use of genetics, have homologues in vertebrates (Dressler & Gruss, 1988). In certain cases, the evolutionary conservation of developmental genes between insects and vertebrates seems to reflect a conservation of function. The best example of this is the family of Antennapedia-like homeobox genes which seem likely to have analogous roles in specifying position along the body in insects and vertebrates (Holland & Hogan, 1988; Graham *et al.*, 1989). However, for certain other conserved genes, their roles in insects and vertebrates seem not to be directly related, and this is illustrated below by discussion of the proto-oncogene *int*-1.

Genes found in other biological contexts

A number of genes initially implicated in specific biological processes in contexts other than embryos have expression patterns suggestive of rather different roles during normal development. For example, certain genes associated with the growth of cultured cells, or with tumour formation, seem to be normally involved in the induction, organisation or differentiation of tissues rather than in cell growth. In part, this may reflect the utilisation of genes for distinct functions in different tissues, and it is likely that many more developmental genes will be discovered in this manner.

Analysis of developmental expression patterns

In situ hybridisation has been the most widely used method for analysing developmental expression patterns in the mouse, as specific probes can be rapidly obtained from cDNA or genomic clones. The expression pattern yields, at the very least, important information regarding the tissue localisation of transcripts that is essential for planning and interpreting manipulative experiments. The nature of the sites of transcript accumulation may argue against certain ideas as to developmental function; for example, expression in post-mitotic cells argues against a role only in cell proliferation. Moreover, correlations between the expression pattern and developmental events, or analogies with patterns observed in other systems, can suggest ideas as to what specific process the gene may be involved in. It must be emphasised that the latter ideas are to be regarded as highly tentative, but can suggest specific approaches towards testing function. Below, the developmental expression patterns of the proto-oncogene *int*-1 and the growth factor-regulated gene *Krox*-20 will be discussed to illustrate the ways in which *in situ* hybridisation can be used to suggest potential function.

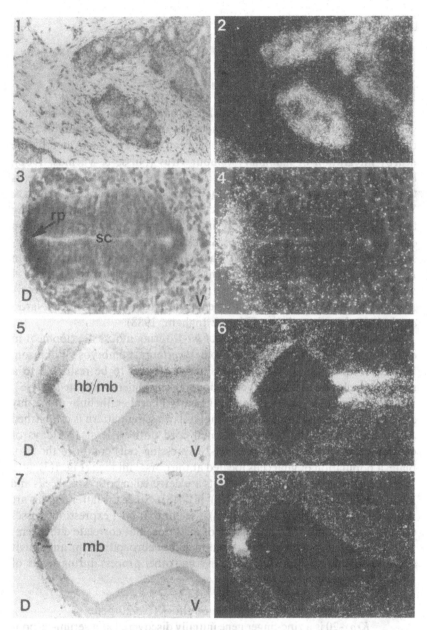

Figure 1. The expression of *int*-1 in mammary tumours and embryos. *int*-1 expression was revealed by *in situ* hybridisation of RNA probe to sections, followed by washing and autoradiography. 1, 2; section through *int*-1-expressing mammary tumour. 3–8; transverse sections through the spinal cord (3, 4), hindbrain/midbrain (5, 6) and midbrain (7, 8) of an 11.5 day mouse embryo. 1, 3, 5 and 7 are bright-field photographs corresponding to the dark-field images in 2, 4, 6 and 8. D, dorsal, V, ventral; rp, roof plate; sc, spinal cord; hb, hindbrain; mb, midbrain. (See *Cell*, **50**, 79–88 for further examples of the expression pattern.)

Results

Developmental expression of the proto-oncogene int-1

The *int*-1 gene was initially identified as a proto-oncogene associated with the formation of mammary tumours in the mouse. In many of these tumours, the *int*-1 gene is found to have been abberantly activated by the nearby integration of mammary tumour virus (Nusse & Varmus, 1982), leading to expression in epithelial cells (Fig. 1(1), (2)). The *int*-1 gene appears to encode a secreted protein product (Fung *et al.*, 1985; Papkoff, Brown & Varmus, 1987) and has a highly restricted normal expression pattern, being found by RNA blot analysis only in adult testes and in mid-gestation embryos (Jakobovits *et al.*, 1986; Shackleford & Varmus, 1987). Subsequently, it was found that *int*-1 has a highly conserved homologue in *Drosophila*, and that this is the 'segment polarity' gene *wingless* (Rijsewijk *et al.*, 1987). The latter gene is involved in the patterning of segments in the early *Drosophila* embryo, and appears to mediate cell–cell interactions (DiNardo *et al.*, 1988; Martinez-Arias, Baker & Ingham, 1988).

We studied the expression pattern of *int*-1 during mouse development in order to gain clues as to its normal role in vertebrate embryos (Wilkinson, Bailes & McMahon, 1987b). Expression was found to be restricted to a specific population of cells in the developing central nervous system (CNS). *int*-1 transcripts are first detected in the early neural epithelium in 8.5 day embryos, and shortly thereafter a complex expression pattern is established and maintained in the brain and spinal cord until at least 14.5 days of development. In the spinal cord, *int*-1-expressing cells comprise the roof plate, a population of glial cells located at the dorsal midline (Fig. 1(3), (4)). In the developing brain *int*-1 transcripts are detected in both dorsal cells and in lateral and ventral regions (Fig. 1(5)–(6)–1(8)). Since the latter regions are neuronal (rather than exclusively glial) it seems that *int*-1 expression crosses cell-type boundaries. The expression of *int*-1 does not correlate either temporarily or spatially with cell proliferation in the neuroepithelium, and thus it seems likely that this gene has a role in some other process during stages of CNS development in the mouse.

Developmental expression of the *Krox*-20 gene

Krox-20 is a zinc-finger gene initially discovered as a serum-responsive gene in NIH3T3 fibroblasts (Almendral *et al.*, 1988; Chavrier *et al.*, 1988). *Krox*-20 transcripts are rapidly and transiently induced on adding serum or purified growth factors to quiescent fibroblasts. Thus, since this gene potentially encodes a transcription factor, it was believed that *Krox*-20 may have a very early role in gene regulation during the resumption of cell growth.

In situ hybridisation analysis of *Krox*-20 expression during mouse development revealed an unexpected expression pattern in the developing CNS (Wilkinson *et al.*, 1989). In the 8.5 day embryo, transcripts were detected in two patches of cells in the neural epithelium (Fig. 2(1), (2)). These sites of expression are located in the prospective hindbrain, and at 9.5 days of development repeated bulges, termed neuromeres, have formed in this region of the central nervous system. Recent studies (Lumsden and Keynes, 1989) have confirmed long-standing speculations that neuromeres reflect a segmental organisation of the early hindbrain (see Lumsden & Keynes, 1989 and Wilkinson *et al.*, 1989 for references). At 9.5 days of development, *Krox*-20 transcripts are detected in two alternate neuromeres, with sharp boundaries of expression coincident with the neuromeric sulci (Fig. 2(3)–(6)). These neuromeres can be identified as the third and fifth, since the otocyst lies between neuromeres 5 and 6 (Vaage, 1969). By 10 days of development, *Krox*-20 expression is no longer detected in neuromere 3 (Fig. 2(7), (8)), and by 10.5 days expression is barely detected in the hindbrain (not shown). Subsequently, *Krox*-20 transcripts transiently accumulate in two hindbrain nuclei in a pattern seemingly unrelated to the earlier segment-restricted expression (Wilkinson *et al.*, 1989). Although it has been suggested that each neuromere is a centre of proliferative activity in the hindbrain (Bergquist & Kallen, 1954), mitotic cells are found in all neuromeres (and the rest of the neuroepithelium). Thus, the expression of *Krox*-20 specifically in neuromeres 3 and 5 argues against a simple association with cell proliferation, unless a segment-specific role is proposed.

Discussion

Potential roles of *int-1* and *Krox*-20 in CNS development

In *Drosophila*, the *int-1*/*wingless* gene forms part of a cascade of interactions that establish and pattern segments in the early *Drosophila* embryo, and its expression is coupled to that of other segmentation genes. *wingless* appears to mediate cell-cell interactions that regulate the expression of other segmentation genes (DiNardo *et al.*, 1988; Martinez-Arias *et al.*, 1988). However, the expression pattern of its murine homologue, *int*-1, is not associated with segmentation, which within the CNS of higher vertebrates is only apparent within the hindbrain. Thus, the specific developmental role of *int*-1/*wingless* is likely to be different in *Drosophila* and mice, and this may be explained by the observation that segmentation seems to have evolved independently in insects and vertebrates (see Clarke, 1964 for discussion and references); it is likely that different sets of genes have been recruited into establishing segments in these two systems (see below). The most reasonable extrapolation that can be made from *Drosophila* to mice is that *int*-1 is likely

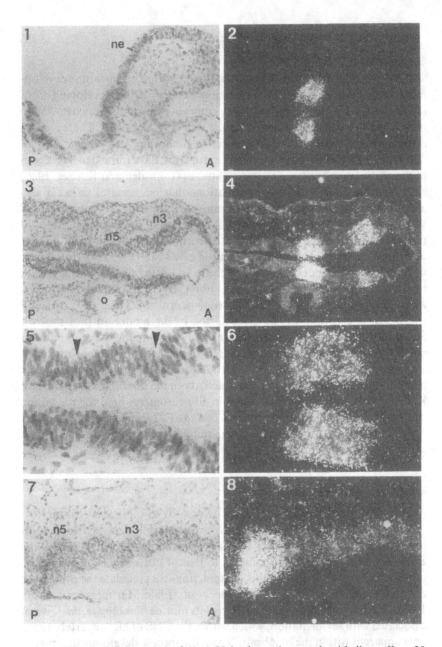

Figure 2. Expression of *Krox*-20 in the early neural epithelium. *Krox*-20 transcripts were detected by *in situ* hybridisation to mouse embryo sections. 1, 2; longitudinal section of 8.5 day embryo. 3, 4; coronal section of 9.5 day embryo. 5, 6; higher magnification view of neuromere 5 in 3, 4. 7, 8; longitudinal section through 10 day embryo. 1, 3, 5 and 7 are bright-field photographs corresponding to the dark-field images shown in 2, 4, 6 and 8. A, anterior; P, posterior; ne, neural epithelium; n3, n5, neuromeres 3 and 5, respectively; o, otocyst. The arrows indicate the neuromeric boundaries. (Modified from *Nature (London)*, **337**, 461–5.)

to be involved in cell-cell interactions. What process might such interactions be involved in? At present, there are few clues as to *int*-1 function in the early CNS, since the developmental role of the *int*-1-expressing roof plate cells is not understood. However, the location of these cells at the dorsal midline is consistent with a potential role in patterning along the dorso-ventral axis of the spinal cord. Conceivably, a gradient of *int*-1 protein may be established that specifies position along this axis. Another possibility is that *int*-1 has a more local role in, for example, organising the tracts of sensory axons that form either side of the roof plate in the spinal cord (Altman & Bayer, 1984).

The expression of *Krox*-20 in alternate hindbrain segments is analogous to that of a class of *Drosophila* segmentation genes: the pair-rule genes. The latter genes have been shown by genetic studies to be involved in the establishment of segments in the early embryo (Nusslein-Volhard & Weischaus, 1980), and it is possible that *Krox*-20 may have an analogous role in the vertebrate hindbrain. Although none of the *Drosophila* pair-rule genes encode proteins with zinc-fingers, several segmentation gap genes do, although these are certainly not homologues of *Krox*-20. The latter considerations may seem inconsistent with *Krox*-20 being a segmentation gene, but as mentioned above this may be explained by the independent evolution of segmentation in insects and vertebrates. We envisage that the establishment of segments by alternating patterns of gene expression may be a developmental strategy that has evolved independently in these lineages, and different genes have been recruited to this process.

The above speculations are largely based on developmental expression patterns and analogies with *Drosophila* embryogenesis, but seem inconsistent with the assumed physiological effects of *int*-1 and *Krox*-20 in their initial sites of discovery. Both of these genes are implicated in cell growth, *int*-1 in mammary tumours and *Krox*-20 in cultured fibroblasts, yet their developmental expression patterns do not correlate with cell proliferation. One possible explanation for this is that these genes are multifunctional, affecting cell growth in certain contexts, but being involved in other processes elsewhere. This senario is analogous to that for multifunctional 'growth factors' which in different contexts can stimulate or repress cell proliferation, or act as morphogens (Mercola & Stiles, 1988). *Krox*-20 expression occurs in the thymus and spleen of adult mice (Chavrier *et al.*, 1988), and also in certain glial cells in the developing peripheral nervous system (Wilkinson *et al.*, 1989) and could be involved in cell proliferation in these tissues. On the other hand, *int*-1 is not normally expressed in mammary cells, and so its role in mammary tumour formation may involve it mimicking an unidentified mitogen in this tissue.

Methods for identifying developmental genes in mice

The above discussion has a number of implications for methods of identifying developmental genes. Although the analysis of vertebrate homologues of *Drosophila* genes may uncover important developmental genes, in some cases they may not have the same specific role in these diverse systems. The most convincing case for a conserved function is for the *Antennapedia*-like homeobox genes which may specify differences in pattern along the body axis, an ancient feature probably shared by the common ancestor of vertebrates and insects. However, a similar function for homologous genes is less likely for those involved in developmental processes that have arisen separately in these lineages. A corrolary of this is that certain developmental genes (such as *Krox*-20) may not be shared by insects and vertebrates. This raises a problem for the analysis of the molecular mechanisms of tissue patterning in vertebrate embryos: understanding of this process will require the identification of all of the relevant genes, but how can we find those that cannot be found through homologies? In the case of hindbrain segmentation, we envisage that this process involves a cascade of interactions between genes with segment-restricted expression, but there is at present no systematic approach for finding such genes. Some mutants may exist that have aberrant hindbrain segmentation (Deol, 1964), but cloning of the corresponding genes is likely to be slow.

One possible approach is to identify by differential screening genes that are specifically expressed in the segmented hindbrain, but not in the non-segmented spinal cord during early CNS development. Certain of these genes may, like *Krox*-20, have segment-restricted expression patterns and these could be identified by *in situ* hybridisation to appropriate sections through neuromeres. However, if *in situ* hybridisation is to be used as a screening method, it is essential that rapid methods exist for producing probes from candidate cDNA clones. Furthermore, the differential screening involves using RNA from tissues that are available in only tiny quantities. These technical problems may be overcome by some recently developed methods. First, cDNA libraries can be constructed in phage vectors with T7 and SP6 RNA polymerase sites, and such that the cDNA is in a defined orientation with respect to these promoters (Palazzolo & Meyerowitz, 1987). Thus, it is possible to produce antisense RNA probes directly from purified cDNA clones, without the need for subcloning or determining the orientation of transcription. Furthermore, such libraries can be used for the amplification of RNA for use in differential screening. Second, strategies exist for the amplification of small amounts of cDNA by the use of polymerase chain reactions (Frohmann, Dush & Martin, 1988; Loh *et al*., 1989; Belyavsky, Vinogradova & Rajewsky, 1989) and thus libraries can be derived from small amounts of microdissected tissues. It is anticipated that these methods will allow the rapid identification of RNAs with spatially restricted expression that can be further characterised by *in situ* hybridisation analysis.

Protocols

In situ hybridisation was carried out by the method of Cox *et al.* (1984) with the modifications described previously (Wilkinson *et al.*, 1987a, b). An important feature of this protocol is that high stringency washing was carried out following hybridisation in order to achieve low non-specific backgrounds. The method is described briefly below.

Fixation, embedding and sectioning of embryos

Embryos were fixed in 4% paraformaldehyde in PBS overnight at 4°C, and then successively incubated for at least 30 min in 0.85% saline (twice, at 4°C), 50% ethanol/50% saline, 70%, 70%, 85%, 95%, 100% and 100% ethanol. Next, the embryos were transferred through toluene (twice for 30 min), 50% toluene/50% wax at 60°C (20 min) and paraffin wax (three changes of 20 min each at 60°C using Fibrowax, BDH). The embryos were orientated in a glass embryo dish under a dissection microscope before allowing the wax to set. 6um sections were cut on a microtome, and mounted on slides subbed with 3-aminopropylethoxysilane (Rentrop *et al.*, 1986).

Pretreatment, hybridisation and washing of sections

Wax was removed from mounted sections by two 10 min incubations in xylene, the sections rehydrated through an ethanol series, and postfixed in 4% paraformaldehyde in PBS for 20 min. Following this, the sections were treated for 5 min with 10 μg/ml proteinase K, the postfixation was repeated, and the slides were treated for 10 min with 0.25% acetic anhydride in 0.1 M triethanolamine. Finally, the slides were dehydrated through an ethanol series and dried. Hybridisation under coverslips using ^{35}S-labelled anti-sense RNA probes (2×10^5 cts/min/μl) was carried out at 55°C overnight in a moist chamber. Next, the coverslips were removed in $5 \times$ SSC, 10 mM DTT, the slides subjected to a high stringency wash in 50% formamide, $2 \times$ SSC, 10 mM DTT at 65°C for 30 min. The slides were then washed in 0.5 M NaCl, 10 mM Tris HCl, pH 8.0, 10 mH EDTA prior to incubation at 37°C for 30 min with 20 μg/ul ribonuclease A in the latter buffer. The high stringency washing was repeated once more and then the slides were dehydrated through an ethanol series and air-dried.

Autoradiography

The slides were dipped in 1:1 diluted Ilford K5 nuclear track emulsion at 43°C, and dried slowly at room temperature for 4 h before setting them up for exposure at 4°C. Typically, slides were exposed for ~6 days prior to developing in Kodak D19 developer (2 min), incubation in 1% acetic acid, 1% glycerol (1 min) and fixing in 30% sodium thiosulphate (2 min). The sections were stained with toluidine blue and mounted using Permount (Fisher Scientific).

References

Akam, M. E. (1987). The molecular basis for metameric pattern in *Drosophila* embryos. *Development*, **101**, 1–22.

Almendral, J. M., Somer, D., Macdonald-Bravo, H., Burckhardt, J., Perera, J. & Bravo, R. (1988). Complexity of the early genetic response to growth factors in mouse fibroblasts. *Molecular Cellular Biology*, **8**, 2140–8.

Altman, J. & Bayer, S. S. (1984). The *Development of the Rat Spinal Cord*. Berlin, Heidelberg, New York, Tokyo: Springer-Verlag.

Belyavsky, A., Vinogradova, T. & Rajewsky, K. (1989). PCR-based cDNA library construction: general cDNA libraries at the level of a few cells. *Nucleic Acids Research*, **17**, 2919–32.

Bergquist, H. & Kallen, B. (1954). Notes on the early histogenesis and morphogenesis of the central nervous system in vertebrates. *Journal Comparative Neurology*, **100**, 627–60.

Chavrier, P., Zerial, M., Lemaire, P., Almendral, J., Bravo, R. & Charnay, P. (1988). A gene encoding a protein with zinc fingers is activated during G_0/G_1 transition in cultured cells. *The EMBO Journal*, **7**, 29–35.

Clarke, R. B. (1964). *Dynamics in Metazoan Evolution: The Origin of the Coelom and Segments*. Oxford: Oxford University Press.

Cox, K. H., Deleon, D. V., Angerer, L. M. & Angerer, R. C. (1984). Detection of mRNAs in sea urchin embryos by *in situ* hybridisation using asymmetric RNA probes. *Developmental Biology*, **101**, 485–502.

Deol, M. S. (1964). The abnormalities of the inner ear in *kreisler* mice. *Journal of Embryological Experimental Morphology*, **72**, 475–90

Dressler, G. R. & Gruss, P. (1988). Do multigene families regulate vertebrate development? *Trends in Genetics*, **4**, 214–19.

DiNardo, S., Sher, S., Heemsterk-Jongers, J., Kassiss, J. A. & O'Farrell, P. H. (1988). Two-tiered regulation of spatially patterned engrailed gene expression during *Drosophila* development. *Nature (London)*, **332**, 604–9.

Frohman, M. A., Dush, M. K. & Martin, G. R. (1988). Rapid production of full-length cDNAs from rare transcripts: Amplification using a single gene-specific oligonucleotide primer. *Proceedings of the National Academy of Sciences, USA*, **85**, 8998–9002.

Fung, Y. K. T., Shackleford, G. M., Brown, A. M. C., Sanders, G. S. & Varmus, H. E. (1985). Nucleotide sequence and expression in vitro of cDNA derived from mRNA of *int*-1, a provirally activated mouse mammary oncogene. *Molecular Cellular Biology*, **5**, 3337–44.

Graham, A., Papalopulu, N. & Krumlauf, R. (1989). The murine and

Drosophila homeobox gene complexes have common features of organisation and expression. *Cell*, **57**, 367–78.
Holland, P. W. H. & Hogan, B. L. M. (1988). Expression of homeobox genes during mouse development: a review. *Genes Development*, **2**, 773–82.
Ingham, P. W. (1988). The molecular genetics of embryonic pattern formation in *Drosophila*. *Nature (London)*, **335**, 25–34.
Jakobovits, A., Shackleford, G. M., Varmus, H. E. & Martin, G. R. (1986). Two proto-oncogenes implicated in mammary carcinogenensis, *int*-1 and *int*-2, are independently regulated during mouse development. *Proceedings of the National Academy of Sciences, USA*, **83**, 7806–10.
Loh, E. Y., Elliott, J. F., Cwirla, S., Lanier, L. L. & Davis, M. M. (1989). Polymerase chain reaction with single-sided specificity: analysis of T cell receptor delta chain. *Science*, **243**, 217–20.
Lumsden, A. & Keynes, R. (1989). Segmental patterns of neuronal development in the chick hindbrain. *Nature (London)*, **337**, 424–8.
Martinez-Arias, A., Baker, N. E. & Ingham, P. W. (1988). Role of segment polarity genes in the definition of cell states in the *Drosophila* embryo. *Development*, **103**, 157–70.
Mercola, M. & Stiles, C. D. (1988). Growth factor superfamilies and mammalian embryogenesis. *Development*, **102**, 451–60.
Nusse, R. & Varmus, H. (1982). Many tumors induced by mouse mammary tumor virus contain a provirus integrated in the same region of the host genome. *Cell*, **31**, 99–109.
Nusslein-Volhard, C. & Weischaus, E. (1980). Mutations affecting segment number and polarity in *Drosophila*. *Nature (London)*, **287**, 795–801.
Palazzolo, M. J. & Meyerowitz, E. M. (1987). A family of lambda phage cDNA cloning vectors, SWAJ, allowing the amplification of RNA sequences. *Gene*, **52**, 197–206.
Papkoff, J., Brown, A. N. C. & Varmus, H. E. (1987). The *int*-1 proto-oncogene products are glycoproteins that appear to enter the secretory pathway. *Molecular Cellular Biology*, **7**, 3978–84.
Rentrop, M., Knapp, B., Winter, H. & Schweizer, J. (1986). Aminoalkylsilane-treated glass slides as support for *in situ* hybridisation of keratin cDNAs to frozen sections under varying fixation and pretreatment conditions. *Histochemical Journal*, **18**, 271–6.
Rijsewijk, F., Schuermann, M., Wagenaar, E., Parren, P., Weigel, D & Nusse, R. (1987). The *Drosophila* homologue of the mouse mammary oncogene *int*-1 is identical to the segment polarity gene *wingless*. *Cell*, **50**, 649–57.
Shackleford, G. M. & Varmus, H. E. (1987). Expression of the proto-oncogene *int*-1 is restricted to postmeiotic male germ cells and the neural tube of mid-gestational embryos. *Cell*, **50**, 89–95.
Vaage, S. (1969). the segmentation of the primitive neural tube in chick embryos (*Gallus domesticus*). A morphological, histochemical and autoradiographic investigation. *Advances in Anatomical Embryological*

Cellular Biology, **41**, 1–88.

Wilkinson, D. G., Bailes, J. A., Champion, J. E. & McMahon, A. P. (1987a). A molecular analysis of mouse development from 8 to 10 days *post coitum* detects changes only in embryonic globin expression. *Development*, **99**, 493–500.

Wilkinson, D. G., Bailes, J. A. & McMahon, A. P. (1987b). Expression of the proto-oncogene *int*-1 is restricted to specific neural cells in the developing mouse embryo. *Cell*, **50**, 79–88.

Wilkinson, D. G., Bhatt, S., Chavrier, P., Bravo, R. & Charnay, P. (1989). Segment-specific expression of a zinc-finger gene in the developing nervous system of the mouse. *Nature (London)*, **337**, 461–5.

Geoffrey Ian McFadden

Evolution of algal plastids from eukaryotic endosymbionts

Introduction

Endosymbiosis has been a key process in the evolution of eukaryotes. Molecular phylogenetic data prove 'beyond reasonable doubt' that mitochondria and green chloroplasts are of eubacterial origin (Pace *et al.*, 1986), probably derived from endosymbionts (Margulis, 1981). Hence, the eukaryotic cell is a polyphyletic chimera of two or more organisms created by cross-bridges between different streams of evolution. Accordingly, the plant cell comprises three subcellular compartments (cytoplasm, chloroplast and mitochondrion) derived from three distantly related organisms now amalgamated into a single cell (Margulis, 1981). However, the morphology of certain algal cells (e.g. the Divisions (or Phyla) Cryptoophyta and Chlorarachniophyta) suggests that they contain four different subcellular compartments; possibly derived from a union of two eukaryotes (Whatley & Whatley, 1981, Cavalier-Smith 1988, Cattolicco, 1986, Cavalier-Smith & Lee, 1985). In order to determine the affinity of the sub-cellular compartments in these algae, high resolution *in situ* hybridisation using lineage specific probes for rRNA has been employed. This cytochemical approach demonstrates that cryptomonads and *Chlorarachnion* not only have two prokaryotic subcelluar compartments (plastids and mitochondria) but also contain two eukaryotic subcellular compartments, each with 80S-like eukaryotic ribosomes and a nucleus. These results support the hypothesis that certain algae attained their plastids from eukaryotic endosymbionts, and that vestigial remnants of the endosymbionts nucleus and cytoplasm remain in the algal cells (Whatley & Whatley, 1981; Hibberd & Norris, 1984; Ludwig and Gibbs, 1985, 1987, 1989; Hansmann, 1988).

Cryptomonads

Cryptomonads are small phytoflagellates found in marine and freshwater habitats. The cells usually have a single plastid containing chlorophylls a and c plus phycobiliprotein pigments (Gantt, 1980). In contrast to the chloroplasts of higher plants and green algae, the plastids in Cryptomonads are bounded by four membranes, an inner pair of envelope membranes and an outer pair of membranes known as the chloroplast endoplasmic reticulum (CER) which are continuous with the nuclear envelope (Gibbs, 1981). On the inner side of one plastid the two pairs of membranes diverge, thereby creating a small compartment known as the periplastidal space (Fig. 1). Within the periplastidal space are particles resembling 80S ribosomes (Fig. 5). In addition, there is an organelle that resembles a eukaryotic nucleus (Figs 2 and 5). The morphology of this organelle lead to it being called a 'nucleomorph' (Greenwood, 1974), but since this chapter and previous works (Ludwig & Gibbs, 1985, 1987; Hansmann et al., 1985, 1986; Hansmann, 1988) have demonstrated that the structure has many characteristics typical of eukaryotic nuclei, I propose that it be renamed the miniature nucleus. The miniature nucleus is delimited by a double membrane with small pores (Figs 1 and 5), and has been shown cytochemically to contain DNA (Ludwig & Gibbs, 1985, 1987; Hansmann et al., 1985, 1986) as well as RNA within a small nucleolus-like region (Hansmann, 1988).

Polynucleotide probes specific for prokaryotic and eukaryotic rRNA label different subcellular compartments in cryptomonad cells. The eukaryotic probe labels the cytoplasm and also the nucleolus of the main nucleus (Fig. 3). The prokaryotic probe labels the plastid (Fig. 4) and, to a lesser extent, the mitochondrion (not illustrated).

Figure 6 shows a view of the plastid, periplastidal space, and miniature nucleus of a cryptomonad cell labelled with the probe for prokaryotic rRNA detected with colloidal gold. The chloroplast ribosomes in the stroma are heavily labelled. The chloroplast ribosomes in the cytoplasm are not labelled. The ribosomes in the periplastidal space are not labelled. The nucleolus-like body of the miniature nucleus is also devoid of label.

Figure 7 shows a similar view to Figure 6 but is labelled with a probe for eukaryotic rRNA. The ribosomes of the cytoplasm are labelled but the heterologous ribosomes in the mitochondria and plastid are not labelled. The ribosomes of the periplastidal space and the nucleolus-like body of the miniature nucleus are labelled.

Chlorarachnion

Chlorarachnion is a photosynthetic amoeba (Figs 8 and 9) with plastids containing chlorophylls a and b (Hibberd & Norris, 1984). The morphology of the plastid-associated compartments is apparently analogous to the cryptomonads. The plastids have a double envelope and the thylakoids associated in pairs (Fig. 11). As in cryptomonads, there are two extra membranes termed the CER surrounding the plastid (Figs 10 and 11). However, the CER of *Chlorarachnion* is not directly connected to the nuclear

Endosymbiotic origin of algal plastids

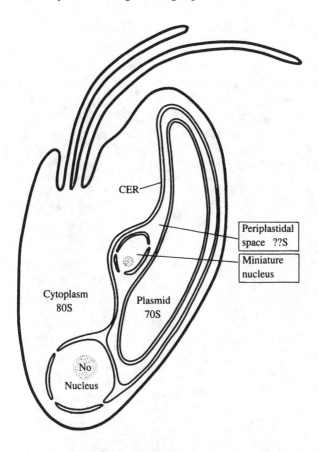

Figure 1. Diagram showing subcellular compartments in a cryptomonad. At the anterior is a gullet with two emergent flagella. The main nucleus with its nucleolus (No) is at the posterior. Four membranes surround the plastid. An extension of the nuclear membranes, the chloroplast endoplasmic reticulum (CER), envelopes the plastid and associated compartment. Between the CER and the plastid envelopes is the periplastidal space which contains ribosomes of unknown affinity. Also within this compartment is the miniature nucleus.

envelope, bears no ribosomes, and may not be entire (Hibberd & Norris, 1984; Ludwig & Gibbs, 1987, 1989). Between the plastid envelopes and CER is a subcellular compartment (periplastidal space) containing ribosome-like particles and a miniature nucleus (Figs 10 and 11). The miniature nucleus of *Chlorarachnion* contains DNA (Ludwig & Gibbs, 1987, 1989) and is delimited by a double membrane which has gaps or pores (Hibberd & Norris, 1984; Ludwig & Gibbs 1989). The miniature nucleus of *Chlorarachnion* is wedge-

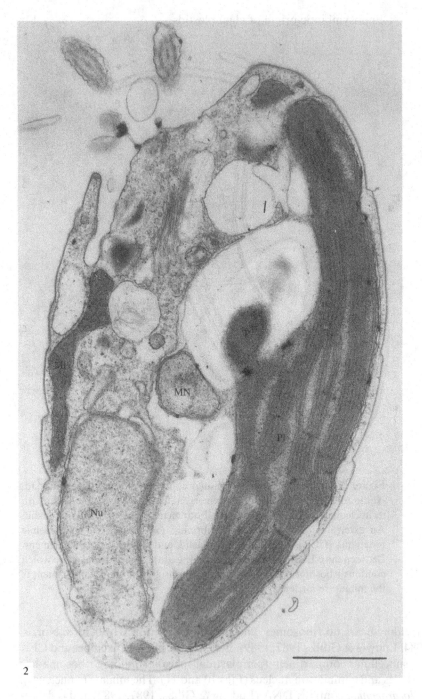

Figure 2. Median longitudinal section of a cryptomonad cell showing the nucleus (Nu), plastid (Pl), pyrenoid (P), mitochondrion (Mi), and the miniature nucleus (MN). (Scale bar = 1 μm.)

Endosymbiotic origin of algal plastids

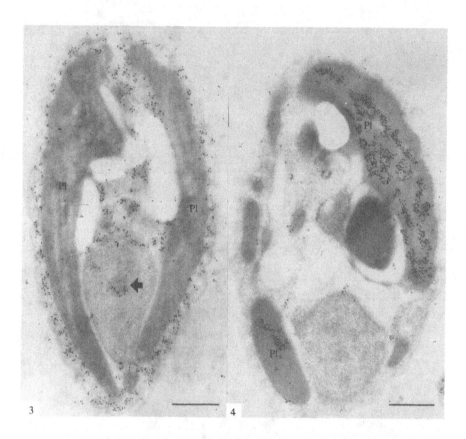

Figure 3. Longitudinal section of cryptomonad cell labelled with probe for eukaryotic rRNA. The cytoplasm and nucleolus (arrow) of the main cell nucleus are heavily labelled with the colloidal gold marker. The plastid (Pl) is not labelled. (Scale bar = 1 μm.)

Figure 4. Longitudinal section of cryptomonad cell labelled with prokaryotic rRNA probe. Only the plastid (Pl) is labelled. (Scale bar = 1 μm.)

shaped and lies within in a notch in the pyrenoid (cf Figs 8, 12 & 13). In *Chlorarachnion* the nucleolus-like region in the miniature nucleus is elongate (Ludwig & Gibbs, 1989).

When hybridised with the eukaryotic probe, the amoebal cytoplasm and nucleolus of the main nucleus in *Chlorachnion* is labelled (Fig. 12). In addition the nucleolus-like region of the miniature nucleus is also labelled (Figs 12–14). When hybridised with the prokaryotic specific rRNA probe, only the ribosomes in the chloroplast and mitochondrion are labelled (not illustrated).

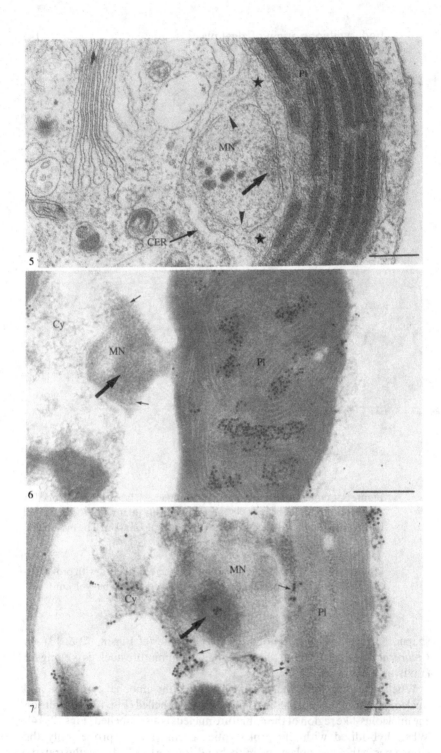

Endosymbiotic origin of algal plastids 149

Discussion

High resolution *in situ* hybridisation shows that the ribosomes in the periplastidal space of cryptomonads and *Chlorarachnion* have more homology to eukaryotic rRNA than prokaryotic rRNA. The periplastidal space is thus a second, separate eukaryotic compartment within these cells. This is congruent with the possibility that the periplastidal space is a vestigal remnant of the cytoplasm of a eukaryotic endosymbiont (Ludwig & Gibbs 1989, Hansmann, 1988).

Where are the genes for periplastidal rRNAs? The CER of cryptomonads has no gaps or pores (Gibbs, 1981, McFadden unpublished) meaning that the ribosomes of the periplastidal compartment are physically isolated; the CER separates them from the main nucleus and cytoplasm (Fig. 1). In the absence of any known mechanism for transport of rRNA across membranes, it must be assumed that the genes for periplastidal rRNA reside in the periplastidal space. Since the *in situ* hybridisation results demonstrate the presence of eukaryotic rRNAs in the nucleolus-like structure of the miniature nucleus, it is highly probable that these are transcripts from periplastidal rRNA genes aggregated into a nucleolus in the miniature nucleus. Presumably the rRNAs exit the miniature nucleus via the pores in the double envelope. The presence of a functional nucleolus in the miniature nucleus of cryptomonads is further demonstration of the eukaryotic nature of this organelle. Eukaryotic rRNA transcripts were also localised to the nucleolus-like region of the miniature nucleus of *Chlorarachnion*, suggesting that genes for periplastidal ribosomes of *Chlorarachnion* are also harboured within the miniature nucleus.

Why do these cells maintain a separate eukaryotic compartment in the periplastidal space? A possible clue lays in the close association of the

Figure 5. Section showing plastid (Pl), CER, periplastidal space (★), and miniature nucleus (MN) with pores (arrowheads) and nucleolus-like region (arrow). (Scale bar = 250 nm.)

Figure 6. Similar view to Figure 5 labelled with the probe for prokaryotic rRNA. The colloidal gold label is concentrated over the ribosomes in the plastid stroma. The ribosomes in the cytoplasm (Cy) and periplastidal space (small arrows) are not labelled. The nucleolus-like region of the miniature nucleus (large arrow) is also devoid of label. (Scale bar = 250 nm.)

Figure 7. Plastid region labelled with eukaryotic rRNA probe. The plastid (Pl) is not labelled. The ribosomes of the cytoplasm (Cy) and periplastidal space (small arrows) are labelled. Within the miniature nucleus (MN) the nucleolus-like region (large arrow) is labelled. (Scale bar = 250 nm.)

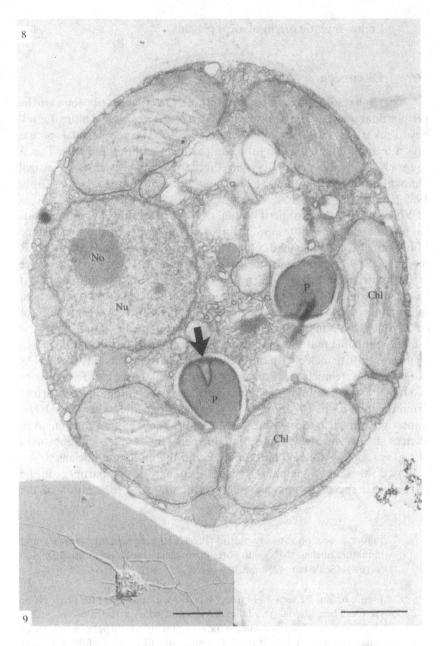

Figure 8. Section of *Chlorarachnion* showing main nucleus (Nu) and nucleolus (No), chloroplasts (Chl) and with bulbous pyrenoids (P), and a miniature nucleus positioned in a cleft of a pyrenoid (arrow). (Scale bar = 1 μm.)

Figure 9. Light micrograph (differential interference contrast) of *Chlorarachnion* amoeba showing reticulopodia. (Scale bar = 1 μm.)

Endosymbiotic origin of algal plastids 151

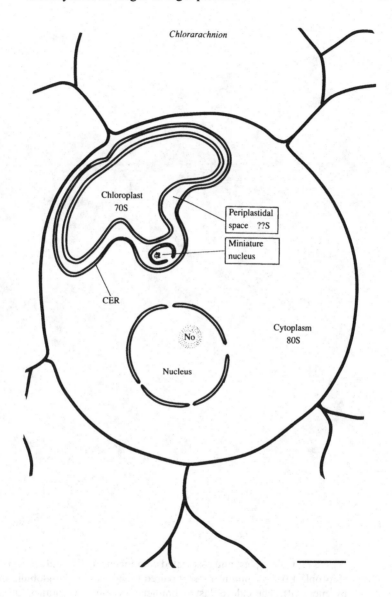

Figure 10. Diagram showing subcellular compartments in *Chlorarachnion*. The chloroplast is bounded by four membranes: the two chloroplast envelopes and the CER. Between the two pairs of membranes is the periplastidal space and the miniature nucleus. (Scale bar = 1 μm.)

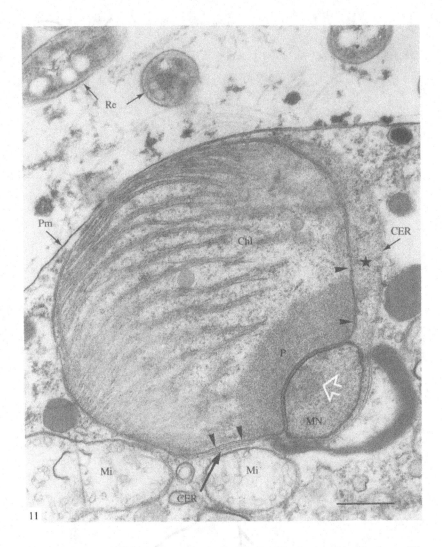

Figure 11. Chloroplast and associated components in *Chlorarachnion*. The chloroplast (Chl) contains loosely paired thylakoids, plastoglobuli, and a pyrenoid (P). The chloroplast is bounded by four membranes: the two envelope membranes (arrowheads) and the CER. Between these pairs of membranes is the periplastidal space (★). The miniature nucleus (MN) lies within the periplastidal space adjacent the pyrenoid. The double membrane of the miniature nucleus and the chloroplast envelopes run parallel in the cleft of the pyrenoid. The faintly staining nucleolus-like region of the miniature nucleus is indicated by a white arrow. Two mitochondrial profiles (Mi), with tubular cristae, lie adjacent the chloroplast. The plasmamembrane (Pm) and reticulopodia (Re) are also visible. (Scale bar = 400 nm.)

miniature nucleus and periplastidal cytoplasm with the plastid. If any plastid proteins are encoded by DNA in the miniature nucleus, then perhaps the cells need to retain the miniature nucleus and its translation system (periplastidal ribosomes) to synthesise chloroplast proteins which are then translocated across the plastid envelopes into the plastid.

While the *in situ* hybridisation results show that the periplastidal compartments in cryptomonads and *Chlorarachnion* are fundamentally eukaryotic, they do not provide any indication of the relatedness of the periplastidal compartments to the main eukaryotic cytoplasm. It is possible that the miniature nucleus and periplastidal ribosomes are derived from the main nucleus and cytoplasm, and are merely partitioned off into the periplastidal compartment by the CER. If cryptomonads and *Chlorarachnion* contain eukaryotic endosymbionts, then the nucleotide sequence of cytoplasmic and periplastidal rRNAs should be divergent. I am currently cloning and sequencing rRNA genes from cryptomonads and *Chlorarachnion* to determine the affinities of the periplastidal compartments.

Materials and methods

Algae

Batch cultures of the cryptomonad (*Chroomonas caudata* Geitler MUCC Cr10) and *Chlorarachnion reptans* Geitler (UTEX LB2314) were maintained as described (Hibberd & Norris, 1984).

Specimen preparation

Cells are fixed for standard electron microscopy with 1% glutaraldehyde and 1% OsO_4 in 50 mM PIPES (pH7) on ice for 15 min. The fixative buffers for *Chlorarachnion* (which is grown in saline media) also contained 0.25 M sucrose. Fixed cells were washed in buffer, dehydrated in a graded ethanol series, and embedded in Spurr's resin. For *in situ* hybridisation the cells were fixed in 2% glutaraldehyde in 50 mM PIPES (pH 7) on ice for 60 min, washed in buffer, then dehydrated through an ethanol series to 70% ethanol and embedded in LR Gold resin (McFadden *et al.*, 1988). Ultrathin sections were collected on pioloform-coated gold grids. Sections were digested in proteinase K (1 μg/ml in TE) for 15 min at 20 °C then rinsed in water and dried.

Probes

The prokaryotic rRNA probe was a 1 kb EcoRI fragment (Eco 36/37 according to Grant *et al.*, 1980) from the chloroplast genome of *Chlamydomonas reinhardtii* encoding plastid 16S rRNA subcloned into pBluescript (Stratagene). The eukaryotic rRNA probe was a 1 kb BamHI/EcoRI nuclear genomic fragment from *Pisum sativum* internal to the 18S rRNA gene (Jorgensen *et al.*, 1987) subcloned into pGEM (Promega).

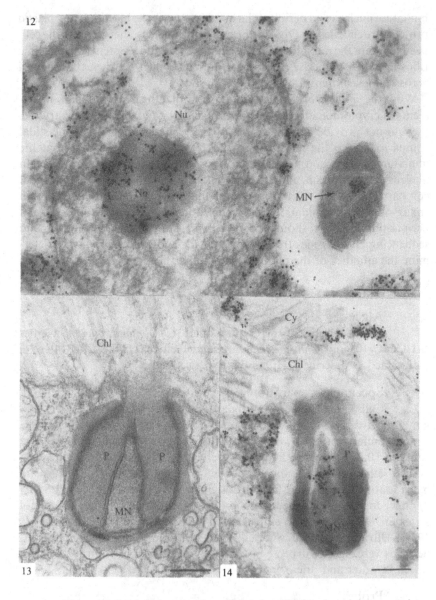

Figure 12. Section showing the main nucleus (Nu) and a miniature nucleus (MN) of *Chlorarachnion* labelled with eukaryotic rRNA probe. The nucleolus of the main nucleus (No) is heavily labelled, as is the nuclear envelope. The nucleoplasm (Nu) is not labelled. In the miniature nucleus, which lies in a cleft of the pyrenoid (cf Fig. 8), only a small area corresponding to the nucleolus-like region is labelled. The pyrenoid matrix (P) is not labelled. (Scale bar = 400 nm.)

Biotinylated sense and antisense RNA probes were polymerised using bio-11-rUTP as described by McFadden (1989).

In situ hybridisation

Grids were inverted onto a 3.5 µl droplet of probe in hybridisation buffer (McFadden et al., 1988) and hybridised. The prokaryotic probe was used at 52 °C and the eukaryotic probe at 60 °C. Grids were rinsed in 4 × SSC, and washed in 1 × SSC at 52 °C for 2 h. Grids were then blocked in 1% BSA in SC buffer (McFadden et al., 1988), incubated with rabbit-anti-biotin followed by goat-anti-rabbit conjugated to colloidal gold as described (McFadden, 1989). The grids were washed in water, dried, and stained with uranyl acetate and lead citrate.

Acknowledgements

During this study I was supported by a Queen Elizabeth II Fellowship. I am particularly indebted to Prof. A. E. Clarke for placing the resources of the Plant Cell Biology Research Centre at my disposal during this work, and for encouraging me to develop this line of research. Dr D. Hill provided the culture of the cryptomonad and Dr W. Thompson and Dr N. Gillham provided the clones from which the probes were derived.

References

Cattolico, R. A. (1986). Chloroplast evolution in algae and land plants. *Trends in Ecology and Evolution*, **1**, 64–7.

Cavalier-Smith, T. (1988). Eukaryotic cell evolution. In *Proceedings of the XIVth International Botanical Congress.* pp. 203–23 (eds Greuter, W. & Zimmer, B.) Koenigstein: Koeltz Scientific Books.

Cavalier-Smith, T. & Lee, J. J. (1985). Protozoa as hosts for endosymbioses, and the conversion of symbionts into organelles. *Journal of Protozoology*, **32**, 376–9.

Gantt, E. (1980). Photosynthetic cryptophytes. In *Phytoflagellates.* (ed. Cox, E.), pp. 381–405. Amsterdam: Elsevier.

Figure 13. Longitudinal section of pyrenoid (P) showing miniature nucleus (MN) situated in a deep cleft. The elongate nucleolus-like region occupies the entire profile of the miniature nucleus when sectioned in this plane. (Scale bar = 300 nm.)

Figure 14. Similar view to Figure 13 labelled with probe for eukaryotic rRNA. The cytoplasm (Cy) and nucleolus-like region in the miniature nucleus are labelled. The pyrenoid (P) and chloroplast (Chl) are not labelled. (Scale bar = 300 nm.)

Gibbs, S. P. (1981). The chloroplast endoplasmic reticulum: structure, function, and evolutionary significance. *International Reviews in Cytology*, **72**, 49–99.

Grant, D. M., Gillham, N. W. & Boynton, J. E. (1980). Inheritance of chloropolast DNA in *Chlamydomonas reinhardtii*. *Proceedings of the National Academy of Sciences, USA*, **77**, 6067–71.

Greenwood, A. D. (1974). The Cryptophyta in relation to phylogeny and photosynthesis. In *8th International Conference of EM* (eds Sanders, J. V. & Goodchild, D. J.), pp. 556–7, AAS, Canberra.

Hansmann, P., Falk, H. & Sitte, P. (1985). DNA in the nucleomorph of *Cryptomonas* demonstrated by DAPI fluorescence. *Zeitschrift Naturforschung*, **40c**, 933–5.

Hansmann, P., Falk, H., Scheer, U. & Sitte, P. (1986). Ultrastructural localization of DNA in two *Cryptomonas* species by use of a monoclonal DNA antibody. *European Journal Cellular Biology*, **42**, 152–60.

Hansmann, P. (1988) Ultrastructural localization of RNA in cryptomonads. *Plastoplasma*, **146**, 81–8.

Hibberd, D. J. & Norris, R. E. (1984). Cytology and ultrastructure of *Chlorarachnion reptans* (Chlorarachniophyta *divisio nova*, Chlorarachniophyceae *classis nova*). *Journal of Phycology*, **20**, 310–30.

Jorgensen, R. A., Cueller, R. E., Thompson, W. F. & Kavanagh, T. A. (1987) Structure and variation in rDNA of pea. Characterization of a cloned repeat and chromosomal rDNA variants. *Plant Molecular Biology*, **8**, 3–12.

Ludwig, M. & Gibbs, S. P. (1985). DNA is present in the nucleomorph of cryptomonads: further evidence that the chloroplast evolved from a eukaryotic endosymbiont. *Protoplasma*, **127**, 9–20.

Ludwig, M. & Gibbs, S. P. (1987). Are the nucleomorphs of cryptomonads and *Chlorarachnion* the vestigial nuclei of eukaryotic endosymbionts? *Annals of New York Academy of Science*, **501**, 198–211.

Ludwig, M. & Gibbs, S. P. (1989). Evidence that the nucleomorphs of *Chlorarachnion reptans* (Chloroarachniophyta) are vestigial nuclei: Morphology, division and DNA-DAPI fluorescence. *Journal of Phycology*, **25**, in press.

McFadden, G. I,, Cornish, E. C., Bonig, I. & Clarke, A. E. (1988). A simple fixation and embedding method for use in hybridization histochemistry of plant tissues. *Histochemistry Journal*, **20**, 575–86.

McFadden, G. I. (1989). *In situ* hybridization techniques in plant tissues: from macroscopic to ultrastructural resolution. *Cell Biology International Reports*, **13**, 3–21.

Margulis, L. (1981). *Symbiosis in Cell Evolution.* Freeman & Co, San Francisco.

Pace, N. R., Olsen, G. J. & Woese, C. R. (1986). Ribosomal RNA phylogeny and the primary lines of evolutionary descent. *Cell*, **45**, 325–6.

Whatley, J. M. & Whatley, F. R. (1981). Chloroplast evolution. *New Phytologist*, **87**, 233–47.

Susan Y. Wright and Andrew J. Greenland

Localisation of expression of male flower-specific genes from maize by *in situ* hybridisation

Introduction

Flowering in plants is a complex process involving the co-ordinated expression of many genes essential for the differentiation of specialised tissues. Whilst many of these genes may also be expressed in other tissues during growth of the plant, an important group of essential genes will be developmental- and tissue-specific, expressed only at critical stages and in defined tissues during the flowering process. In tobacco flowers for example, there are approximately 10000 ovary-specific and 11000 anther-specific mRNAs representing 40% and 42% of all diverse mRNAs present in the respective organs (Kamalay & Goldberg, 1980, 1984).

We are interested in genes which are expressed specifically in the male flowers of maize. In this monoecious plant the male flowers are borne in the tassel which terminates the main stem and the female flowers on ear branches which arise from nodes midway on the main stem. We have used differential screening of cDNA libraries made from whole tassels to isolate six male flower-specific (MFS) cDNAs from maize. Differential screening has also been used to clone floral-specific cDNAs from tomato (McCormick *et al.*, 1987; Gasser *et al.*, 1989), tobacco (Goldberg, 1988), pollen-specific cDNAs from maize (Stinson *et al.*, 1987) and style-specific cDNAs from *Brassica oleracea* (Nasrallah *et al.*, 1985) and *Nicotiana alata* (Anderson *et al.*, 1985).

Whilst conventional Northern and dot blotting techniques using whole organ RNA preparations can be used to establish the organ specificity of the cDNA clones, it is clear that these techniques do not localise expression to particular tissues within the organ. However, the same DNA or RNA probes used in the latter experiments can also be used in *in situ* hybridisation experiments with organ sections to show the cellular localisation of specific nucleic acid sequences. Although *in situ* hybridisation has been used for some considerable time in animal systems (Hafen *et al.*, 1983; Cox *et al.*, 1984) it is only comparatively recently that these techniques have been adapted for use with plant systems and have contributed enormously to our understanding of tissue-specific expression of plant genes (Cornish *et al.*, 1987; Goldberg, 1988; McFadden *et al.*, 1988; Gasser *et al.*, 1989).

In this article we describe our *in situ* hybridisation methods and illustrate their use in the study of expression of maize MFS genes. As a background we briefly review our work and that of other groups in the isolation of floral-specific genes particularly where gene expression has then been characterised by *in situ* hybridisation.

Isolation and characterisation of MFS cDNAs

We are particularly interested in the period of male flower development in maize associated with meiosis. To clone cDNAs to genes which are expressed in the meiocytes or anther tissues during this period we have constructed two cDNA libraries (Greenland *et al.*, 1989). Library 1 was prepared from poly $[A]^+$ RNA from whole maize tassels bearing early meiotic anthers (most meiocytes in early meiotic prophase) and library 2 from poly $[A]^+$ RNA from whole tassels bearing late meiotic anthers (predominantly diad and early tetrad stages). Figure 1 reviews the library screening procedure used and which yielded five unique early meiotic MFS cDNAs and one unique late meiotic cDNA. Table 1 summarises some of the features of each of these cDNA clones. Expression of the mRNAs of the five MFS cDNAs isolated from the early meiotic library is detected in RNA isolated from both early and late meiotic tassel samples. The mRNAs corresponding to these cDNAs are not wholly specific to male flowers and are detected at considerably lower levels in leaves (pMS10 and pMS18) or in leaves, cobs and roots (pMS1, pMS2 and pMS4) (Table 1). In contrast pMS14 mRNA is found only in late meiotic RNA and is not detected in leaves, cobs or roots (Table 1).

We have examined expression of the genes corresponding to these cDNAs during tassel development using dot blot hybridisations (Fig. 2). The early meiotic mRNAs (pMS1, 2, 4, 10, and 18) accumulate very early in development in tassels less than 2 cm in length (Fig. 4). We have not analysed expression in floral meristems prior to this stage. These mRNAs persist through the meiotic anther stages and then decline as pollen grains mature. In contrast the late meiotic mRNA of pMS14 is not detected in tassels less than 5 cm in length, but increases dramatically as the sporogenous cells of the anther enter meiosis (Fig. 2). The significance of the decline in the level of pMS14 mRNA in 10–15 cm tassels is not understood but clearly merits further examination since it may represent pulses of accumulation of pMS14 mRNA during development. As with the early meiotic mRNAs, pMS14 mRNA declines abruptly as mature pollen accumulates in the anthers (Fig. 2).

These data show that different temporal controls of gene expression occur during development of male flowers in maize. The controls which pro-

Male flower-specific genes from maize 159

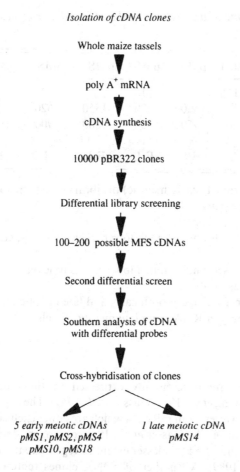

Figure 1. A schematic representation of the differential screening procedure used to isolate MFS cDNAs from libraries prepared from poly [A]$^+$ RNA from whole tassels bearing either early meiotic or late meiotic anthers.

gramme accumulation of the early meiotic mRNAs are probably very similar but contrast markedly with those regulating appearance and accumulation of the late meotic mRNA, pMS14. Both the early and late meiotic mRNAs are involved with developmental processes which occur prior to the accumulation of mature pollen grains. They are clearly not involved with the later stages of anther development such as dehiscence nor are they mRNAs which accumulate in mature pollen.

Differential screening of cDNA libraries prepared from pistils and anthers dissected from tomato flowers has been used to isolate a range of cDNAs

Table 1. *A summary of some of the features of the MFS cDNAs cloned from maize*

	pMS1	pMS2	pMS4	pMS10	pMS14	pMS18
Library[a]	1	1	1	1	2	1
Insert size[b]	750	500	720	1350	620	940
mRNA size[c]	900	950	850	1600	900	1100
Organ specificity[d]	+	+	+	++	+++	++
Expression window[e]	E/L	E/L	E/L	E/L	L	E/L

[a] Isolated from cDNA library 1 (early meiotic) or library 2 (late meiotic).
[b] Approximate size in base pairs.
[c] Approximate size in nucleotides.
[d]   + = expressed in tassels and at much lower levels in leaves, cobs and roots.
   ++ = expressed in tassels and at much lower levels in leaves.
   +++ = expressed in tassels only.
[e] E/L = mRNA present in RNA from both early and late meiotic tassels.
   L = mRNA present only in RNA from late meiotic tassels.

representing genes which are differentially expressed in these organs (McCormick *et al.*, 1987; Wing *et al.*, 1988; Gasser *et al.*, 1989). The degree of organ specificity exhibited varies from mRNAs which are either exclusive to pollen (Wing *et al.*, 1988) or pistils (Gasser *et al.*, 1989) or clones which are expressed in pistils and anthers and to a lesser extent in vegetative tissues of the plant (Gasser *et al.*, 1989). A number of cDNA clones representing anther-specific mRNAs have been isolated from tobacco flowers (Goldberg, 1988). Two of these hybridise to mRNAs which show a similar pattern of expression to maize premeiotic MFS mRNAs in that they accumulate early in anther development and then decay prior to anther dehiscence. In contrast two other tobacco anther mRNAs persist throughout anther and pollen development (Goldberg 1988).

To isolate pollen-specific, gametophytically expressed genes, Stinson *et al.* (1987) differentially screened cDNA libraries made from *Tradescantia* and maize pollen yielding cDNAs which represent mRNAs that accumulate in pollen but not in the surrounding anthers. Unlike maize MFS mRNAs (Fig. 2) and the anther-specific tobacco clones which decline prior to anther dehiscence (Goldberg, 1988), these pollen-specific mRNAs are involved with the later stages of pollen maturation and during germination and pollen tube growth (Mascarenhas, 1988).

Male flower-specific genes from maize

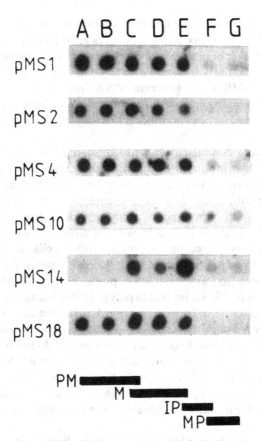

Figure 2. Dot blot analysis of total RNA (4 μg per dot) extracted from maize tassels of increasing length. Total RNA was prepared, bound to nitrocellulose and hybridized to the radiolabelled pMS cDNAS as described (Greenland et al., 1987). All filters were exposed to film for 48 h at −70°C except pMS10 which was exposed for 168 h. The tassel lengths in each sample were as follows: A = <2 cm; B = 2–5 cm; C = 5–10 cm; D = 10–15 cm; E = 15–20 cm; F = 20–30 cm; G = 20–30 cm, shedding pollen. Solid bars show the developmental stage relative to microsporogenesis in each of the samples: **PM**, premeiosis; **M**, meiosis, **IP**, immature pollen; **MP**, mature pollen.

In situ hybridisation with maize MFS cDNAs

One of the primary objectives of our research is to determine the function of the proteins encoded by the MFS genes. We considered that determining the tissue-specificity of expression of the genes by *in situ* hybridisation would be the first step in achieving this objective.

The method we use for *in situ* hybridisation is detailed on pp. 169–71 and

Figure 3 shows examples of the wax embedded sections which are used in these experiments. We find that wax embedding gives good tissue preservation even with the later stages of male flower development where the anthers and microspores are particularly friable. To date we have used two MFS cDNAs, pMS10 and pMS14 to generate RNA probes for *in situ* hybridisation. It is important to emphasise that these methods have not been optimised for either of these probes and this is the subject of our ongoing research.

Our initial studies were conducted with pMS10-derived RNA probes which exhibited strong hybridisation to pollen mother cells of sectioned, early meiotic maize anthers (Fig. 4). However, these data were obtained only with the sense RNA probe and not with antisense probe as would be expected. These data show the importance of using sense and antisense RNA probes and a post-hybridisation treatment with ribonuclease which destroyed the non-specific hybridisation to pollen mother cells (data not shown) but did not affect specific hybrids (see below). Interestingly, in Northern blots both sense and antisense RNA probes from pMS10 bind strongly to residual ribosomal RNA present in poly$[A]^+$ RNA samples from early and late meiotic tassels and from leaves (Fig. 5). The intensity of this non-specific hybridisation is such that it masks specific hybridisation by the antisense probe to pMS10 mRNA. In contrast, DNA probes do not bind to ribosomal subunits, but show specific hybridisation to pMS10 mRNA in the RNA prepared from tassels (Fig. 5). Non-specific binding of RNA probes to ribosomal RNA has been noted by others (Little & Jackson, 1987) and we believe this may account for the false hybridisation signals on anther sections observed here.

Unlike riboprobes derived from pMS10, specific hybridisation to a transcript of the correct size can be detected with an antisense RNA probe derived from pMS14 but not with the sense probe (Fig. 6). As before, some non-specific hybridisation to ribosomal RNA is observed with both sense and antisense probes derived from pMS14. DNA probes do not show background hybridisation and bind specifically to the same 900 nucleotide transcript as the antisense RNA probe (Fig. 6).

Figure 7(a) and (b) show *in situ* hybridisation with pMS14 antisense RNA probes. These hybridisations show that pMS14 mRNA is located in the tapetal cell layer surrounding the developing microspores. Hybridisation of the pMS14 antisense probe does not occur to any other cells in the section. Likewise the pMS14 sense probe does not show any specific hybridisation (Fig. 7(c)). These sections were made from 15–20 cm maize tassels at a stage when the level of pMS14 mRNA is at a maximum (Fig. 2; Greenland *et al.*, 1989). In these sections and in those from subsequent experiments (data not shown) hybridisation occurs to the tapetum of the anthers in one floret but not the other. In Figure 6(a) and (b) the tapetal layers which contain pMS14

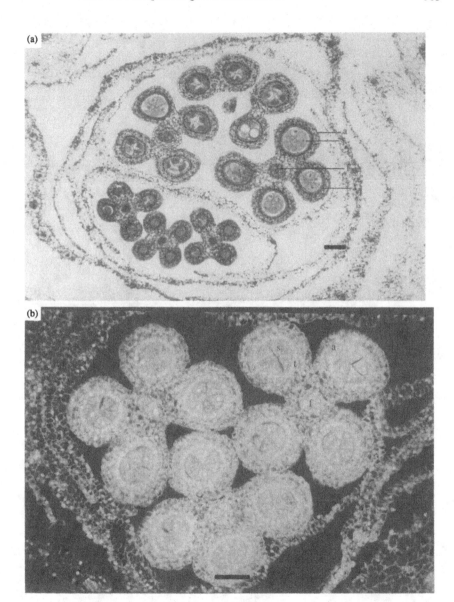

Figure 3. (a) Section through a maize spikelet stained with methyl green pyronin. Note that the anthers in the lower floret within the spikelet contain premeiotic sporogenous cells and the larger anthers in the upper floret contain cells in meiotic prophase. (b) Section through maize anthers in which the sporogenous cells are in meiotic prophase. Stained with acridine orange and photographed under epifluorescence microscopy. **a**, anther; **s**, sporogenous cells; **f**, filament; **t**, tapetum. Scale bars represent 50 μm.

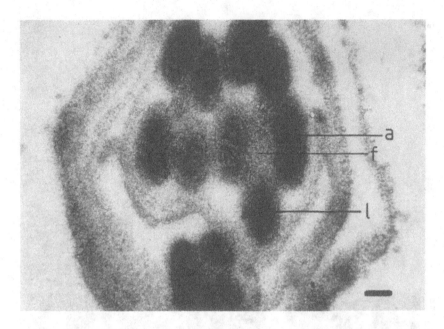

Figure 4. Hybridisation of pMS10-derived sense RNA probes to sections of maize anthers. The hybridisation is non-specific and destroyed by a post-hybridisation treatment with ribonuclease. l, loculus; a, anther; f, filament. Scale bar represents 50 μm.

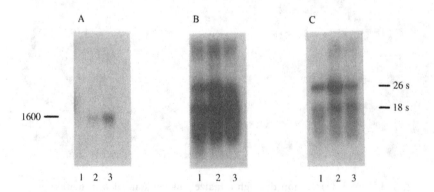

Figure 5. Northern blot analysis of poly [A]$^+$ RNA (1 μg per track) isolated from leaves (1), tassels bearing early meiotic anthers (2) or late meiotic anthers (3) with DNA and RNA probes derived from pMS10. RNA was prepared, separated by denaturing gel electrophoresis and transferred to

Male flower-specific genes from maize

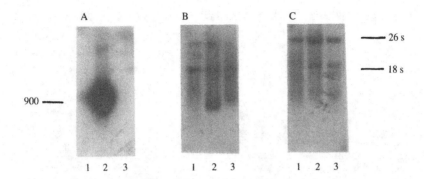

Figure 6. Northern blot analysis of poly [A]$^+$ RNA (1 µg per track) isolated from tassels bearing early meiotic anthers (1) or late meiotic anthers (2) and leaves (3) with DNA and RNA probes derived from pMS14. RNA was prepared, separated by denaturing gel electrophoresis and transferred to nitrocellulose exactly as described (Greenland et al., 1987). Filters were hybridised with ^{32}P-labelled DNA probe from pMS14 (a), and ^{32}P-labelled antisense (b) and sense (c) pMS14 derived RNA probes. Filters were exposed for 18 h at −70°C. The positions of the 26s and 18s ribosomal subunits are shown, as is the specific hybridisation to the 900 nucleotide pMS14 mRNA in (a) and (b).

mRNA surround late meiotic microspores at the tetrad stage whilst the tapetal layers not containing pMS14 mRNA surround sporogenous cells which have not undergone meiosis. It is a feature of maize that the sets of anthers within the individual florets of the spikelets do not develop co-ordinately. This is clearly shown in the stained section in Figure 3. Thus *in situ* hybridisation shows that accumulation of pMS14 mRNA is tissue-specific and confirm data obtained from dot blot analysis (Fig. 2) that expression of pMS14 mRNA is stage specific as it is first detected in the tapetum surrounding meiotic cells.

In situ hybridisation has been used by others to show that anther-specific genes isolated from tomato (Smith et al., 1987) and tobacco (Goldberg, 1988) are expressed in the tapetum of these plants. In the tobacco studies a tapetal-specific mRNA was one of several localised by *in situ* hybridisation

nitrocellulose exactly as described (Greenland et al., 1987). Filters were hybridised with a ^{32}P-labelled DNA probe from pMS10 (a), and ^{32}P-labelled sense (b) and antisense (c) RNA probes. Exposure times at −70°C were 4 h for the RNA probes and 48 h for the DNA probe. The positions of the 26s and 18s ribosomal subunits are shown as is the 1600 nucleotide pMS10 mRNA in (a).

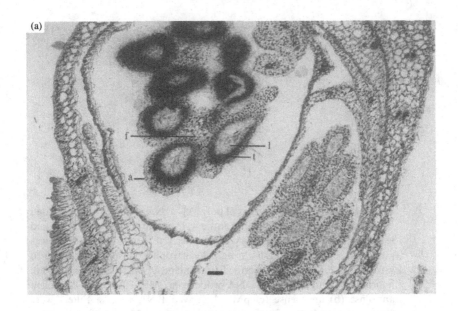

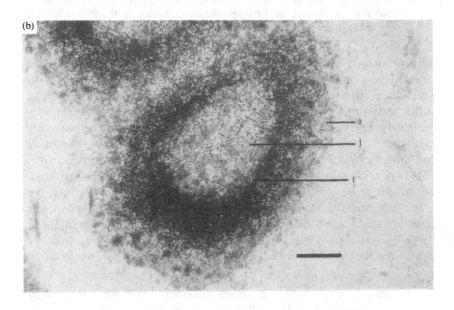

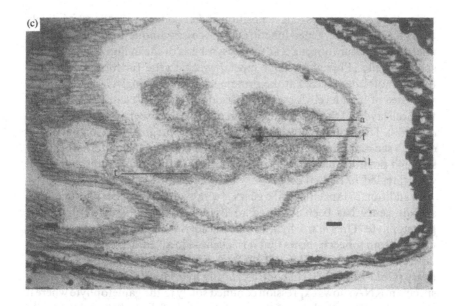

Figure 7. *In situ* hybridisation of maize spikelet sections with single-strand ^{35}S-labelled RNA probes derived from pM514. (a) Hybridisation with antisense probe, (b) higher magnification of previous section, (c) hybridisation with sense probe. **a**, anther; **l**, locule; **t**, tapetum; **f**, filament. Note that the anthers in the lower floret in (a) are at an earlier development stage and do not hybridise to the antisense probe (see text). Scale bars represent 25 μm.

(Goldberg, 1988). The decay of this mRNA parallels the disintegration of the tapetum during the later stages of anther and pollen development (Goldberg, 1988). Similarly it is likely that the degeneration of the tapetum that occurs during the later stages of anther development in maize (Kiesselbach, 1980) accounts for the rapid decline in pMS14 mRNA at this time (Fig. 2). However, unlike pMS14 mRNA which is not detected in premeiotic maize anthers, the tobacco tapetal specific mRNA has been shown by developmental dot blot analysis to accumulate to appreciable levels very early in anther development (Goldberg, 1988).

The nature and function of genes expressed specifically in the tapetum is unknown. In addition, it is not known whether the genes isolated from maize (Greenland *et al.*, 1989) tomato (Smith *et al.*, 1987), and tobacco (Goldberg, 1988) are related. The nucleotide sequence of pMS14 does not exhibit homology with any sequence lodged in the main databases. The tapetum acts as a feeder layer for the developing microspores and is also responsible for the formation of the outer pollen wall or exine. However, since many of the metabolic processes which occur during microspore maturation are also common to other tissues (e.g. deposition and dissolution of callose) it is likely that tapetal-specific genes function in processes unique to that tissue and microspore maturation such as the biosynthesis of the major exine polymer, sporopollenin. Fine analysis of tapetal-specific gene function will require further molecular studies at the protein level. Interestingly, a gene which is expressed at high levels for a short period of time and exclusively in two to three cell layers of the inner integument of ovules has been isolated (Gasser et al., 1989). The function of the inner layers of the integument is to nourish the embryo sac and clear parallels can be drawn between the pattern of expression and possible function of genes expressed in these cells and those which are expressed in tapetal cells.

In addition to tapetal-specific genes, expression of a number of other floral specific genes has been localised to tissues of both the sporophyte and gametophyte. Thus *in situ* hybridisation studies with another anther-specific mRNA from tobacco shows that it is localised in all anther tissues except the tapetum and filament. This mRNA is also present in tissues of the tobacco ovary (Goldberg, 1988). In contrast, *in situ* hybridisation of a maize pollen-specific mRNA shows expression confined solely to the gametophyte where it is located in the cytoplasm of the vegetative cell of the pollen and is distributed throughout pollen tube cytoplasm after germination (Hanson et al., 1989).

Conclusions

In situ hybridisation has been successfully used by a number of groups to localise expression of mRNAs to the tissues of the anther and to pollen. We have used this technique to show that pMS14 mRNA is both developmental stage- and tissue-specific. In our hands *in situ* hybridisation has worked well with single strand RNA probes prepared from a cDNA which represents a relatively abundant mRNA within the poly $[A]^+$ RNA fraction of maize tassels bearing late meiotic anthers. Expression of pMS14 mRNA is confined to the tapetal cell layer yet the corresponding cDNA can be detected at a frequency of 0.1% in libraries made from whole tassels. Since the latter are complex organs comprised of many tissues it is likely that pMS14 mRNA is extremely abundant within tapetal cells. Further experiments are required to estimate the relative abundance of pMS14 mRNA by RNAase protection assays (Little & Jackson, 1987). *In situ* hybridisation using RNA probes from pMS10, which based on frequency of occurrence within cDNA libraries is at least 10-fold less abundant, have so far proved unsuccessful but have underlined the importance of including the appropriate controls in the experiments. Our current work in this area concentrates on optimising conditions for *in situ* hybridisation with maize anther sections, allowing the tissue localisation of further MFS mRNAs to be examined.

Protocols

In situ hybridisation procedure for maize tassel sections

Solutions
1. DEPC H_2O. Add 0.1% diethyl pyrocarbonate (DEPC) to double distilled H_2O, shake vigorously, leave overnight at room temperature and then autoclave. *Note:* all solutions should be made with DEPC H_2O. Wherever possible try to use new, disposable plasticware or treat glassware with 0.1% DEPC overnight and then autoclave.
2. Paraformaldehyde (4%). Add 4 g paraformaldehyde to 30 ml DEPC H_2O. Heat gently and add 1 M NaOH dropwise until the solution clears. Add 2.5 g sucrose (RNAase free) and 50 ml 0.1 M Na phosphate buffer pH 7.2. Make volume up to 100 ml and add Silwet L-77 to 0.005%.
3. Hybridisation buffer is 50% deionised formamide, 1 × Denhardts solution (0.02% bovine serum albumin, 0.02% Ficoll 400,000, 0.02% polyvinyl pyrrolidone 40000), 0.01 M sodium phosphate pH 7.2, 0.3 M NaCl, 0.05 mg/ml denatured salmon sperm DNA, 0.01 M dithiothreitol, 10% dextran sulphate.
4. Formamide wash buffer is 50% formamide, 1 × Denhardts solution, 0.01 M sodium phosphate pH 7.2, 0.3 M NaCl, 0.01 M Tris.HCl pH 7.2, 0.005 M EDTA.

Fixing, embedding and sectioning
1. Vacuum infiltrate small tassel pieces (< 5 mm) or individual spikelets in 4% paraformaldehyde. After 4–6 h transfer to fresh solution and leave overnight with slow shaking.
2. Dehydrate the samples exactly as described (Johansen, 1940). Your samples will finish in 2-methyl propan-2-ol:paraffin oil (50:50 vol/vol).
3. Pour tissue onto solid fibrowax in a glass universal maintained at 40 °C on a water bath. Decant off as much of the 2-methyl propan-2-ol:paraffin oil mixture as possible. Turn water bath up to 60 °C so that tissue sample slowly sinks into the melting wax. Leave the universal cap loose allowing excess 2-methyl propan-2-ol to evaporate. After 1 h transfer sample to fresh

molten wax and leave at 60°C overnight. The following day transfer sample again and leave for a further 5 h. Finally embed the sample in pre-warmed cube trays adding molten wax as necessary. When all the cubes are filled place the tray on ice to set quickly.
4. Sub enough slides for your experiment as described in Rentrop et al., (1986).
5. Cut sections from wax embedded material using a microtome. We find that 12μm sections of maize spikelets give the best preservation of structure particularly in the locules of late meiotic anthers. However, thinner sections (5–10 μm) can be taken from less friable material. Sections are floated out on warm (40°C) water and caught on subbed slides. The slides are drained, dried at 37°C overnight and stored at room temperature until required.

In situ hybridisation

1. Dewax and proteinase K treat your sections on subbed slides as follows:

Solution	Time (min)	Temperature
xylene	2 × 5	RT[a]
xylene/ethanol[b]	5	RT
100% ethanol	5	RT
70% ethanol	2	RT
50% ethanol	2	RT
30% ethanol	2	RT
DEPC H_2O	2	RT
proteinase K[c]	30	37°C
DEPC H_2O	1	RT
4% paraformaldehyde	20	RT
DEPC H_2O	1	RT
30% ethanol	1	RT
50% ethanol	1	RT
70% ethanol	1	RT
100% ethanol	1	RT

[a] RT = room temperature.
[b] 50:50 by volume
[c] 1 μg/ml in 0.05 M Tris.HCl pH 7.5, 0.005 M EDTA, dripped onto each slide.

Finally air dry the slides at room temperature. If required as a negative control, an RNAase treatment (0.5 mg/ml in 0.2 M Na phosphate buffer pH 6.5 for 30 min at 37°C) and H_2O rinse (1 min at RT) may be included immediately after the proteinase K step. Keep these slides separate and take precautions to ensure that your experimental slides do not become contaminated with RNAase.

2. Add 50–75 μl of hybridisation buffer containing your probe (approx 10^7 cts/min) to each slide and cover with a coverslip. Place the slides on a suitable flat tray (e.g. lid of a large petri or bioassay dish) and cover with clingfilm. Seal in a large plastic box lined with damp paper towels. Incubate overnight at 50°C without shaking. We use sense and antisense [^{35}S] labelled RNA transcripts as probes. These are produced from small (< 600bp) fragments of cDNA clones which have been subcloned into pBS vectors (Stratagene, La Jolla, USA). The labelling reactions with T3 and T7 polymerases are done following instructions provided by the suppliers of the plasmid.

3. After hybridisation very carefully remove the coverslips and wash the slides in a plastic box as follows:
    – 4 × 1 h in 200 ml of formamide wash buffer at 50°C with gentle shaking.
    – 1 × 30 min in 20 μg/ml RNAase A, 0.01 M Tris.HCl pH 8, 0.001 M EDTA, 0.5 M NaCl at 37°C without shaking.
    – 4 × 15 min in 250 ml of 2 × SSC at 65°C with gentle shaking.
    – 2 × 15 min in 250 ml of 0.1 × SSC at 65°C with gentle shaking.

4. After washing, drain the slides and dehydrate in a series through 30%, 60% and 85% ethanol containing 0.3 M ammonium acetate, 2 min each step. Dry the slides at room temperature.

5. Pipette film emulsion onto each slide and drain off excess. Allow to set with slide horizontal. Store in light-proof container at 4°C.

6. Develop slides for 5 min in Ilford Phenisol, fix for 5 min in standard fixer (e.g. Agfa G353) and wash in water for 30 min. Drain the slides, allow to dry, and mount in glycerol under a coverslip.

Acknowledgements

We are grateful to Cliff Hart, ICI Agrochemicals for his help with fixation and sectioning of plant material and for photomicroscopy. We would also like to acknowledge Dave Mow, ICI Agrochemicals for preparation of photographs for publication.

References

Anderson, M. A., Cornish, E. C., Mau, S. C., Williams, E. G., Hoggart, R., Atkinson, A., Bonig, I., Grego, B., Simpson, R., Roche, P. J., Haley, J. D., Penschow, J. D., Niall, H. D., Tregear, G. N., Coghlan, J. P., Crawford, R. J. & Clarke, A. E. (1985). Cloning of cDNA for a stylar glycoprotein associated with expression of self-incompatibility in *Nicotiana alata*. *Nature (London)*, **321**, 38–44.

Cornish, E. C., Pettitt, J. M., Bonig, I. & Clarke, A. E. (1987). Developmentally controlled expression of a gene associated with self incompatibility in *Nicotiana alata*. *Nature (London)*, **326**, 99–102.

Cox, K. H., DeLeon, D. V., Angerer, L. M. & Angerer, R. C. (1984). Detection of mRNAs in sea urchin embryos by *in situ* hybridization using asymmetric RNA probes. *Developmental Biology*, **101**, 485–502.

Gasser, C., Budelier, K. A., Smith, A. G., Shah, D. M. & Fraley, R. T. (1989). Isolation of tissue-specific cDNAs from tomato pistils. *The Plant Cell*, **1**, 15–24.

Goldberg, R. B. (1988). Plants: Novel developmental processes. *Science*, **240**, 1460–7.

Greenland, A. J., Bell, P. J., Suner, M-M., Vaudin, M. & Wright, S. Y. (1989). Isolation and characterization of male flower-specific cDNAs from maize, in press.

Greenland, A. J., Thomas, M. V. & Walden, R. M. (1987). Expression of two nuclear genes encoding chloroplast proteins during early development of cucumber seedlings. *Planta*, **170**, 99–110.

Hafen, E., Levine, M., Garber, R. L. & Gehring, W. J. (1983). An improved *in situ* hybridization method for the detection of cellular RNAs in *Drosophila* tissue sections and its application for localizing transcripts of the homeotic *Antennapedia* gene complex. *The EMBO Journal*, **2**, 617–23.

Hanson, D. D., Hamilton, D. A., Travis, J. L., Bashe, D. M. & Mascarenhas, J. P. (1989). Characterization of a pollen-specific cDNA clone from *Zea mays* and its expression. *The Plant Cell*, **1**, 173–9.

Johansen, D. A. (1940). *Plant Microtechnique*. New York: McGraw-Hill Book Co.

Kamalay, J. C. & Goldberg, R. B. (1980). Regulation of structural gene expression in tobacco. *Cell*, **19**, 935–46.

Kamalay, J. C. & Goldberg, R. B. (1984). Organ-specific nuclear RNAs in tobacco. *Proceedings of the National Academy of Sciences, USA*, **81**, 2801–5.

Kiesselbach, T. A. (1980). *The Structure and Reproduction of Corn.* Lincoln: University of Nebraska Press.

Little, F. R. & Jackson, I. J. (1987). Application of plasmids containing promoters specific for phage-encoded RNA polymerases. In *DNA Cloning: A Practical Approach*, vol. III (ed. D. M. Glover), pp. 1–18. Oxford: IRL Press.

Mascarenhas, J. P. (1988). Anther- and pollen-expressed genes. In *Temporal and Spatial Regulation of Plant Genes* (eds Verma, D. P. S. & Goldberg, R. B.), pp. 97–115. Vienna: Springer-Verlag.

McCormick, S., Smith, A., Gasser, C., Sachs, K., Hinchee, M., Horsch, R. & Fraley, R. (1987). Identification of genes specifically expressed in reproductive organs of tomato. In *Tomato Biotechnology* (eds Niven, D. J. & Jones, R. A.), pp. 255–65. New York: Alan R. Liss, Inc.

McFadden, G. I., Ahluwalia, B., Clarke, A. E. & Fincher, G. B. (1988). Expression sites and developmental regulation of genes encoding 1-3, 1-4-beta glucanases in germinated barley. *Planta*, **173**, 500–8.

Nasrallah, J. B., Kao, T-H., Goldberg, M. C. & Nasrallah, M. E. (1985). A complementary cDNA clone encoding an S-locus-specific glycoprotein from *Brassica oleracea*. *Nature (London)*, **318**, 263–7.

Rentrop, M., Knapp, B., Winter, H. & Schweizer, J. (1986). Amino-alkylsilane-treated glass slides as support for *in situ* hybridization of keratin cDNAs to frozen tissue sections under varying fixation and pretreatment conditions. *Histochemistry Journal*, **18**, 271–6.

Smith, A. G., Hinchee, M. & Horsch R. (1987). Cell and tissue specific expression localized by *in situ* hybridisation in floral tissues. *Plant Molecular Biology Reporter*, **5**, 237–41.

Stinson, J. R., Eisenberg, A. J., Willing, R. P., Pe, M. E., Ganson, D. D. & Mascarenhas, J. P. (1987). Genes expressed in the male gametophyte of flowering plants and their isolation. *Plant Physiology*, **83**, 442–7.

Wing, R., Larabell, S., Yamaguchi, J. & McCormick, S. (1988). Molecular characterization of pollen specific cDNA clones from tomato. *Journal of Cellular Biochemistry*, Supplement 12C, 168.

N. Harris, J. Mulcrone & H. Grindley

Tissue preparation techniques for *in situ* hybridisation studies of storage-protein gene expression during pea seed development

Introduction

The values of the *in situ* hybridisation technique, and a background to probes and hybridisation conditions have been discussed elsewhere in this volume. This chapter describes the application of the technique to the study of seed development and considers some different approaches to specimen preparations for histological, cytological and ultrastructural localisation of mRNAs.

Seeds and fruits represent the major food source for humans, and also provide a fascinating range of 'model' systems for studying development and differentiation in plant tissues (e.g. reviews Goldberg, 1988; Goldberg *et al.*, 1989). During seed and fruit development, specific genes, such as the storage protein genes, are expressed to high levels and exhibit a substantial degree of temporal and spatial regulation. In developing pea (*Pisum sativum* L.) seeds only three multigene families, consisting of approximately 20 genes, give approximately 80% of the total protein synthesis. These genes encode the storage proteins vicilin, a glycosylated 7S, legumin, a non-glycosylated 11S, and con-vicilin, a 7S-type protein.

The general morphological features of the development of dicot seeds are well established, and there have been several detailed reviews of the biochemistry and molecular biology of both embryogenesis (Dure, 1985) and seed storage proteins (e.g. Higgins, 1984; Gatehouse *et al.*, 1986). Seed development progresses in a series of well-defined phases following fertilisation. Initial cell division of the zygote results in a 'globular' stage embryo. With subsequent assymetric divisions the cotyledons and plumular apex begin to develop, giving rise to the characteristic 'heart-shaped' stage, although there is very little enlargement of cells; the increase in embryo size is a result of the 'phase of cell division'. During this stage of fruit development a coenocytic endosperm is still present. In the 'monocot' cereals, such as barley and wheat, the endosperm tissue cellularises and becomes the major storage tissue. In a few dicots, such as tobacco and castor bean, a cellularised endosperm develops and acts as a major storage tissue, while the embryo, which maintains a 'heart' or 'torpedo' shape, contains only the minor fraction of the seed reserves. In pea however, as in many dicots, the endosperm has only a transient presence before being crushed between the enlarging embryo and the embryo sac boundary wall. When the 'phase of cell division' is complete, the embryo enlarges as a result of progressive expansion of the cotyledonary cells. It is during the phase of cell expansion that the storage reserves are laid down. In pea these are predominantly starch, which is accumulated within the plastids, and protein which is synthesised at the endoplasmic reticulum, transported via the Golgi and accumulated within the membrane-bound protein bodies (review Chrispeels, 1985). During the phase of expansion and storage deposition there is little, if any, division of the established storage parenchyma cells although there is some continued division within the developing vascular elements. Expansion of the storage parenchyma cells is initially accompanied by expansion of the central vacuole at which time there is synthesis of stored starch but not protein. With the onset of the synthesis of large amounts of storage protein, the central vacuole is progressively replaced by a population of smaller vacuoles and vesicles. These develop, in the still-expanding cells, into the protein bodies (Harris & Boulter, 1976; Craig et al., 1979). Following the phase of expansion and storage deposition there is a phase of 'seed maturation' during which metabolic activities and moisture content drop dramatically; there may also be synthesis and accumulation of 'stress' proteins prior to seed dispersal.

Molecular data on the regulation of storage protein gene expression has come from various approaches including solution and filter hybridisation studies (e.g. Kamalay & Goldberg, 1980), correlations of levels of protein synthesis with extant levels of specific mRNAs (e.g. Croy et al., 1988), and analysis of the expression of, for example, the effects of deletion constructs (Shirsat et al., 1989a) and gene copy number (Shirsat et al., 1989b) on transgenic expression of a pea legumin storage protein in transgenic tobacco.

Along with others (e.g. Barker et al., 1988), we have utilised *in situ* hybridisation studies to link our molecular and structural work on developing seeds (Harris & Croy, 1986), and also correlated *in situ* localisation of mRNAs with immunocytochemical studies of protein distribution (Harris et al., 1989). Because of the rapid development in embryo size and complexity, and also our interest in developing techniques for histological, cytological and ultrastructural localisations, a range of techniques have been employed. These have included the use of both radioactively-labelled and biotinylated probes, and their use in systems involving both pre- and post-embedding hybridisations. These different approaches are outlined on pages 183–5, and their appropriateness to the study of the different stages of embryo development is discussed in pages 180–3.

Early embryo development

Cooper (1938) and Reeve (1948) established the main cytological and morphological events of early embryo development in pea, whilst Marinos (1970) detailed the structural relationship between the embryo, endosperm and seed coat. In contrast to events during animal embryogenesis, there is no cell migration or translocation during the growth of plant embryos. The transition from a globular to heart-shaped embryo is the result of differential cell division and expansion in the various regions of the embryo. Cell polarisation at the start of embryogenesis has been reported (e.g. Raghaven, 1981) but the nature of the controlling factors of embryo development, for example, transcriptional, translational and/or hormonal, is still largely unknown. Presently various laboratories are screening cDNA libraries from different stages of embryo development in attempts to identify controlling elements, initially of unknown function. This has been used successfully in various studies of developmental botany, for example see Wright et al. and Dickinson et al. (this volume), but an alternative approach, that of following back the temporal and spatial patterns of expression of known, organ and tissue-specific genes, has also proved useful in studies of the control of development and differentiation (e.g. Nelson & Langdale, 1989; McFadden, 1989). Our studies have included examining the spatial and temporal patterns of storage protein gene expression during embryo development.

Because of the practical problems of preparing sufficient slides of very young embryos *in situ*, we first examined young embryos isolated from ovules. Our preliminary experiments indicated that it is possible to carry out *in situ* hybridisation with whole young embryos up to 1 to 2 mm in diameter (i.e. up to the globular/heart-shaped transition). The embryos were first fixed and pretreated to allow subsequent permeation of the tissues with the probe and reagents. Biotinylated probes are appropriate with fluorescent visualisation. Sample size is large, in relation to sections, and this requires prolonged washing to ensure complete removal of unbound probe and also stain. The resultant specimen, however, shows distribution of the bound probe in a three-dimensional manner within the various layers of the complete embryo. Using a conventional epipolarising system it is possible to achieve a crude degree of optical sectioning but the subsequent photography is always disappointing; the application of confocal should however overcome such problems (see Ingham; this volume). Using this whole-mount approach, without a confocal system, we have found that the very young embryo exhibits low levels of storage protein mRNAs. Legumin mRNA has been demonstrated in both the globular and heart-shaped embryos. At these stages of development most cells are non-vacuolate, contain relatively low amounts of endoplasmic reticulum, and are dominated by the nucleus. Fluorescent staining showed legumin mRNA located in all cells throughout the embryo except those of the epidermal layer of the plumular apex; the epidermal cells of the developing cotyledons and radicle were stained. The fluorescence was relatively uniform throughout the cytosol, but such a staining protocol with later stages resulted in discrete foci of labelling (see below). Because of the general dogma that storage protein synthesis is associated with the phase of cell expansion later in embryo development we were somewhat surprised to find low levels of legumin mRNA in these early embryos. However, low levels of legumin have been detected at embryo stages prior to cessation of cell division (Domoney & Casey, 1984). The nature of an experimental technique using whole mounts has the obvious potential for problems of incomplete removal of unbound probe or dye, and we were particularly careful with stringency and post-staining washes; the consistent presence of the unlabelled plumular epidermal cells also acted as a form of internal control.

The whole mount approach is, however, not suitable for studying the embryo within the ovule, and for such work we have used both *in situ* hybridisation and immunocytochemistry on sections of previously fixed and embedded material. The protocol for tissue preparation is outlined on pages 184–5; we have found that embedding in the water-soluble medium PEG (polyethylene glycol) is preferable to wax with regard to retention of detail of tissue structure, although the sections are somewhat harder to handle. Both radioactively labelled and biotinylated probes have been used for *in situ* hybridisation with sections from the PEG-embedded tissue.

The low level of legumin mRNA in the very young globular embryos was confirmed with [^{35}S] labelled probes on tissue sections; a very high level of labelling of the coenocytic endosperm surrounding the embryo and lining the embryo-sac boundary wall was also demonstrated, although the large fluid-filled void in the middle of the ovule was unlabelled (Harris *et al.*, 1989).

With the exception of its absence from the plumular epidermal cells, we found no evidence of any patterns of differential expression of legumin mRNA during the progressive development of the heart-shaped embryo and during its subsequent expansion prior to the onset of high levels of legumin mRNA synthesis.

Recent ELISA assays have indicated that low levels of storage protein are present in the very young ovules (Mulcrone, unpublished observations).

Immunocytochemical examination of sections equivalent to those demonstrating legumin mRNA in the endosperm has indicated that there is also a low level accumulation of legumin protein in the endosperm, although no significant immunolabelling was found associated with the globular and heart-shaped embryos, despite the presence of low levels of legumin mRNA. The presence of both legumin mRNA and protein in the endosperm is transient; neither message nor protein were found in endosperm in ovules with the more mature embryos at the phase of cell expansion.

During the early stages of embryo development the suspensor cells are a prominent feature; the suspensor cells expand pushing the young embryo away from the micropylar region towards the apex of the ovule, and closer to the end of the main ovular vascular supply (Marinos, 1970; Hardham, 1976). The suspensor cells, which through evolutionary comparisons and transfer cell morphology, have been thought to have a role in embryo nutrition, showed labelling with probes to legumin mRNA but little the storage protein.

Cotyledonary storage deposition

All of the major organs are established during the phase of cell division, and it is the cotyledons that undergo the most dramatic changes in size and composition during subsequent development. Within the expanding cotyledons, the storage parenchyma cells enlarge and show major changes in their endomembrane components (review Harris, 1986). Cotyledons are modified leaves and their expansion is accompanied by development of a vascular network; this involves continued cell divisions and the differentiation of vascular parenchyma, and xylem and phloem initials. The vascular network is known to have a major influence on the mobilisation of storage reserves (e.g. Bewley & Black, 1985), although its relationship to the pattern of storage deposition is less clear.

Little is known about the temporal and spatial patterns of gene expression associated with storage deposition, although a molecular characterisation has been undertaken. During pea seed development, concurrent with the onset of the 'phase of cell expansion', there is a dramatic increase, initially quadratic and subsequently linear, in extant levels of mRNA for both vicilin and legumin (Croy et al., 1988). This is correlated with rapidly changing rates of run-off transcripts from nuclei isolated from appropriately aged tissues (Evans et al., 1979). There is, however, approximately 24 hours between the increase in vicilin mRNA levels and those from the legumin genes (Gatehouse et al., 1986).

Barker et al. (1988) have shown, by *in situ* hybridisation, that in soybean and transformed *Nicotiana*, the storage protein mRNAs are located predominantly in the cotyledons of the developing embryos. In soybean they noted initiation by a 'wave' of mRNA synthesis moving through the cotyledons, and considered this activity a 'switching-on' event rather than an 'upregulation'. However, our results (Harris & Croy, 1986; Harris et al.,

1989) and those of Hauxwell *et al.* (1989) appear to show a different pattern for the onset of storage protein deposition in pea. The major increases in legumin and vicilin mRNAs occur in a relatively synchronous manner throughout the cotyledon, although with approximately 24 hours between the enhancement of vicilin levels and the subsequent enhancement of legumin mRNA. Using embryo weight rather than days after flowering as a finer division of development, Hauxwell *et al.* (1989) have shown preliminary data of the onset of vicilin accumulation occurring throughout the cotyledon in scattered individual cells. The enhanced expression occurs slightly faster in the adaxial parenchyma, and they correlated this pattern of enhanced mRNA levels with that of an earlier reduction in mitotic activity in this tissue.

Throughout the period of deposition storage protein mRNAs remain high within the cotyledonary parenchyma cells, although within the developing vascular network and epidermal cells only low levels of these mRNAs occur. It is only in the latter part of the deposition phase, in the few days prior to the onset of maturation, that enhancement of storage protein mRNAs is seen in these vascular tissues where, as with the storage paranchyma, elevated levels of vacilin preceed legumin. These increases are against an overall drop in vicilin levels within the whole cotyledons although, in the cultivar of pea (Feltham First) that we have studied, the legumin levels remain relatively high until seed maturation.

Comparison of methods

Our results have been obtained from 'whole-mounts' or tissue sections examined by optical microscopy after hybridisation with either ^{35}S-labelled or photobiotinylated probes. For the latter we have used both anti-biotin / alkaline phosphatase and (strept)avidin / marker systems (detailed elsewhere in this volume). Streptavidin-gold and silver enhancement, with epipolarising illumination, has been particularly successful for identifying non-abundant levels of mRNA; this method has an advantage over darkfield illumination of autoradiography preps because small starch grains are clearly distinguishable from silver grains under epipolarised light.

In an attempt to avoid the relatively time-consuming processes of dehydration and embedding of plant tissues we examined both cryosectioning and hybridisation in thin but 'self-supporting' tissue pieces obtained by tissue chopping. Our attempts at cryosectioning large samples of highly vacuolate tissues have not been particularly successful, although for some plant tissues, particularly those that are not highly vacuolate, cryosectioning is appropriate (e.g. McFadden, 1989; and refs therein). Hybridisation in 100 to 200 μm sections of fixed, vacuolate tissue, using fluorescence localisation of biotinylated probes, has shown both histological and some sub-cellular detail (e.g. Harris & Croy, 1986). In comparison to the relatively even distribution of message throughout the cytosol in non-expanded parenchyma cells (see above), cells in the later stages of development show a major part of the storage protein mRNA distributed in discreet foci in a pattern which reflects the distribution of the endomembrane system. This is hardly surprising as we know that these messages are translated at the ER, but it does demonstrate the potential of a relatively simple approach to tissue preparation which still indicates some sub-cellular distribution. The thickness of the 'self-supporting' tissue section is dependent upon cell size and degree of cell adhesion, and is a compromise between the need to get probes into the tissue and the requirement for sufficient structural stability to permit hybridisation, washing and staining.

Seed tissue sections have been used for examination of the distribution of signal at ultrastructural levels (see pages 184–5). Hybridised probe has been localised by avidin–peroxidase, with subsequent DAB staining, and by avidin–ferritin. Even after hybridisation the degree of structural preservation was sufficiently good to allow localisation of storage-protein mRNAs associated with transition vesicles and with vesicles associated with the *cis* face of the highly-polarised dictyosomes which are present in cotyledonary parenchyma tissue (Harris & Croy, 1986). Ferritin is, however, rather limited as a marker because of its relatively low contrast unless used with bismuth subnitrate (Ainsworth & Karnovsky, 1972); colloidal gold is a superior marker and has been used for *in situ* hybridisation at the ultrastructural level for both chromosome studies (Hutshinson, 1984) and, in a post-embedding technique, for examination of the abundant ribosomal RNA (see McFadden, this volume). We failed to get good labelling with a pre-embedding hybridisation and staining procedure, although it is not immediately apparent why; the 5 nm colloid size would not seem to be a problem as ferritin was useful.

Ultrastructural localisation of mRNAs may also be achieved by thin sectioning of samples previously examined by optical microscopy. This approach, outlined on pages 184–5, has the major advantage that samples can be screened by optical microscopy before continuing with the more time-consuming EM studies. Using standard (7 to 12 μm) sections on slides for hybridisation, the target is localised by biotinylated probe and visualised by streptavidin gold and silver enhancement. Appropriate sections are selected, embedded in resin and sectioned for electron microscopy. A major consideration in such an experimental approach is the nature of the primary tissue fixation and embedding, since these will influence the quality of the ultrastructural preservation. Aldehyde-based fixatives are thus more appropriate than coaggulative protocols, although the latter may give higher levels of accessible mRNA for hybridisation (see elsewhere this volume). We have found that wax embedding results in rather poor retention of cytological detail with our specimens, although this is not the case with all tissues.

Embedding in polyethylene glycol (either 1000 or 1500 depending on tissue type and, perhaps, the batch of PEG) results in good retention of ultrastructural detail, the *cis* and *trans* faces of ditysomes are for example clearly distinguishable. From an 'aesthetic' standpoint gold with silver enhancement is rather rough and irregular in size by comparison with the uniform labelling characteristic of EM immunocytochemistry. However, the level of contrast available with this staining is such that lower levels of magnification can be used than with ferritin. The latter is suitable for higher resolution work but its usefulness is limited by relatively low contrast.

Conclusions
Storage protein deposition in developing peas

Seed storage proteins represent a set of highly regulated genes which are expressed to very abundant levels during only a short period of the plant life cycle. Both transcriptional and post-transcriptional control of the expression of such genes have been demonstrated (refs in Goldberg *et al.*, 1989), and transgenic expression studies are proving valuable in determining *cis*-acting control sequences. It will be important to determine whether enhanced expression is generally a 'switch-on' event, as in soybean, or an 'up-regulation', as appears to be the case in pea, or whether there are species differences. The γ glutamine synthetase of *Phaseolus* root nodules was thought to be another example of tissue specific expression, but recently Cullimore's group has used sensitive RNAase protection assays to demonstrate that γ-GS induction during nodulation is an quantitative up-regulation rather than a qualitative change (Forde & Cullimore, 1989). The temporal and spatial patterns of enhanced storage protein expression within pea cotyledons appear to be unrelated to the nutrient supply pathway, although there is delayed upregulation in the abaxial epidermal cells which develop as transfer cells. Neither is there any apparent association with the developing vascular network, although enhanced levels of storage protein mRNA accumulate later in parenchyma *within* the network. During the phase of cell expansion and storage deposition there is continued cell division within the vascular network, and the preliminary results of Hauxwell *et al.* (1989) would appear to confirm a predominant temporal (in relation to time since mitosis) rather than a spatial influence. The delayed mRNA synthesis in abaxial epidermal cells may also be a consequence of later mitotic divisions; Hepher *et al.* (1988) have shown that in soybean the cells of the abaxial epidermis of the cotyledons continue mitotic divisions for longer than the underlying cells. Pea legumin expression in the transient, coenocytic endosperm is intriguing as the gene has often been described as 'embryo-specific'. 11S proteins are accumulated in cereal endosperms and in dicots with a retained cellularised endosperm, and they are subsequently mobilised during germination. Transgenic expression of pea legumin in tobacco resulted in accumulation of the pea protein in both embryo and in the cellularised tobacco endosperm, suggesting that the pea legumin controlling sequences are not embryo specific (Croy et al, 1988). Our results suggest that legumin may be acting as a (very) transient reserve in pea endosperm, although in this case its mobilisation is initiated not during seed germination but during embryo development.

In contrast to the storage proteins, rather less is known about the regulation of other embryo-specific genes which are involved in the early stages of embryogenesis; differential screening of cDNA libraries and, perhaps more importantly, the establishment of mutant series (e.g. Meinke, 1986) may lead to probes which will be valuable in determining early pattern development.

Different approaches to tissue handling for *in situ* hybridisation

At present the most common approach to *in situ* hybridisation for multicellular samples involves application of probe to tissue sections mounted on glass slides. The techniques refined by the Angerers, and detailed elsewhere in this volume, have now become routine in many medical, animal and plant applications. The development of new types of probe labelling (see Coulton, this volume) are resulting in protocols of shorter duration, and, where required, potentially higher resolution. Biotinylated probes can be applied to both tissue sections and to whole mount preparations, and when used with a range of visualisation techniques are appropriate to optical and electron microscopy. For determination of three dimensional information within tissues, fluorescent probes and confocal microscopy would seem to have potential. For ultrastructural studies, labelling either within a larger tissue piece or within a 10 μm section with subsequent embedding and sectioning for EM, can be successful. For the latter approach it is important to choose preparation techniques for the initial LM phase which are appropriate to the retention of ultrastructural detail for the subsequent EM observations. Where abundant, for example ribosomal, RNA is the target McFadden (see this volume) has shown that a hybridisation technique with sections of resin-embedded samples can give results that are both scientifically valuable and aesthetically attractive.

Protocols

The four approaches to *in situ* hybridisation utilised in our work have followed the outline protocols listed below. We have found it necessary to modify various components, for example choice of primary fixation, extent of 'blocking', and method of visualisation etc, to optimise signal to background with the various tissues and stages of development. Because of the high levels of endogenous biotin 'blocking' has been particularly important. Coulton (this volume) has described appropriate methods; we have also found that reduction of the formamide concentration during hybridisation also helps to reduce background, although higher temperatures are of course required.

Pre-embedding technique for LM
Tissue fixed (sectioned to approx 200–300 μm if required) buffered paraformaldehyde / glutaraldehyde
Protease treatment (1 μg/ml proteinase K at 37°C for 30 min)
Re-fixed (4% paraformaldehyde for 10 min)
Pre-treatments (hybridisation buffer + non-probe DNA or RNA)
Hybridised with probe (using cDNA; or sense or antisense RNA)
If RNA probe (RNase 10 μg/ml in 2 × SSC for 20 min)
Stringency washes (SSC to either 2 ×, 1 × or 0.5 ×)
Probe localisation
    (strept)avidin–fluorochrome (5–50 ng/ml)
    SSC washing until control 'clean'.

Pre-embedding technique for EM
Tissue fixed (buffered paraformaldehyde / glutaraldehyde) and sectioned to approx 200–300 μm
Protease treatment (1 μg/ml proteinase K at 37°C for 30 min)
Re-fixed (4% paraformaldehyde for 10 min)
Pre-treatments (hybridisation buffer + nonprobe DNA or RNA)
Hybridised with probe (using cDNA; or sense or antisense RNA)
Stringency washes (SSC to either 2 ×, 1 × or 0.5 ×)
Probe localisation (for biotinylated probes)
    (i) (strept)avidin–peroxidase
        DAB / osmication
    (ii) avidin–ferritin
Embedding and sectioning for EM
    acetone dehydration and Spurr resin.

Post-embedding technique for LM
Tissue fixed (buffered paraformaldehyde / glutaraldehyde)
Embedded in wax or PEG 1500 after dehydration
Sectioned to 10 μm and placed on washed / baked / subbed or TESPA treated slides
Pre-treatments (0.2 M HCl for 30 min RT) 2 × SSC at 70°C for 15 min, wash and blot dry, 1 μg/ml proteinase K at 37°C for 30 min,

refix with 4% paraformaldehyde for 10 min, dehydrate and air dry
Hybridised with probe (with cDNA or RNA (under coverslip)) and after hybridisation c/s removed with 4 × SSC
Stringency washes (2 × SSC + 10 µg/ml RNase, to 2 × SSC, 1 × SSC or 0.5 × SSC)
Probe localisation (for biotinylated probes)
   (i) alkaline phosphatase (BRL)
   (ii) streptavidin gold + silver enhancement (with amplification)
or autoradiography of [^{35}S]labelled probes.

Post-embedding technique for EM
Tissue fixed (buffered paraformaldehyde / glutaraldehyde)
Embedded and sectioned for LM (as in III)
Pre-treatments (as in III)
Hybridised with probe (as in III)
Probe localisation
   streptavidin gold + silver enhancement (with amplification)
Re-embedded in Spurr resin and sectioned for EM.

References

Ainsworth, S. K. & Karnovsky, M. J. (1972). An ultrastructural staining method for enhancing the size and electron opacity of ferritin in thin sections. *Journal of Histochemistry and Cytochemistry*, **2**, 225–9.

Barker, S. J., Harada, J. J. & Goldberg R. B. (1988). Cellular localization of soybean storage protein mRNA in transformed tobacco seeds. *Proceedings of the National Academy of Sciences, USA*, **85**, 458–62.

Bewley, J. D. & Black, M. (1985) *Seeds: Physiology of Development and Germination*. Plenum Publ. Corp., New York.

Boulter, D., Evans, M., Ellis, R., Shirsat, A., Gatehouse, J. A. & Croy, R. R. D. (1987). Differential gene expression in the development of *Pisum sativum*. *Plant Physiology and Biochemistry*, **25**, 283–9.

Chrispeels, M. J. (1985). The role of the Golgi apparatus in the transport and post-translational modification of the vacuolar (protein body) proteins. *Oxford Surveys of Plant Molecular and Cell Biology*, **2**, 43–68.

Cooper, D. C. (1938). Embryology of *Pisum sativum*. *Botanical Gazette*, **100**, 123–32.

Craig, S., Goodchild, D. J. & Hardman, A. (1979). Structural aspects of protein accumulation in developing pea cotyledons: qualitative and quantitative changes in parenchyma cell vacuoles. *Australian Journal of Plant Physiology*, **6**, 81–98.

Croy, R. R. D., Evans, I. M., Yarwood, J. Y., Harris, N., Gatehouse, J. A., Shirsat, A., Kang, A., Ellis, J. R., Thompson, A. & Boulter, D. (1988).

Expression of pea legumin sequences in pea, *Nicotiana*, and yeast. *Biochemie Physiologie Pflanzen*, **183**, 183–97.

Domoney, C. & Casey, R. (1984). Storage protein precursor polypeptides in cotyledons of *Pisum sativum* L. Identification of, and isolation of a cDNA clone for, an 80 000 M_r Legumin related polypeptide. *European Journal of Biochemistry*, **139**, 321–7.

Dure, L. III (1985). Embryogenesis and gene expression during seed formation. *Oxford Surveys of Plant Molecular and Cell Biology*, **2**, 179–97.

Evans, I. M., Croy, R. R. D., Hutchinson, P., Boulter, D., Paynbe, P. I. & Gordon, M. E. (1979). Cell free synthesis of some storage protein subunits by polyribosomes and RNA isolated from developing seeds of pea (*Piscus sativum* L.) cotyledons, *Planta*, **144**, 452–62.

Forde, B. G. & Cullimore, J. V. (1989). The molecular biology of glutamine synthesis in higher plants. *Oxford Surveys of Plant Molecular and Cell Biology*, **6**, in press.

Gatehouse, J. A., Evans, I. M., Croy, R. R. D. & Boulter, D. (1986). Differential expression of genes during legume seed development. *Philosophical Transactions of the Royal Society London Series B*, **314**, 367–84.

Goldberg, R. B. (1988). Plants: novel developmental procession. *Science*, **240**, 1460–7.

Goldberg, R. B., Barker, S. J. & Perez-Grau, L. (1989). Regulation of gene expression during plant embryogenesis. *Cell*, **56**, 149–60.

Hardman, A. (1976). Structural aspects of the pathways of nutrient flow to the developing cotyledons of *Pisum sativum*. *Australian Journal of Botany*, **24**, 711–21.

Harris, N. (1986). Organization of the endomembrane system. *Annual Review of Plant Physiology*, **37**, 73–92.

Harris, N. & Boulter, D. (1976). Protein body formation in cotyledons of developing cow pea (*Vigna unguiculata*) seeds. *Annals of Botany*, **40**, 739–44.

Harris, N. & Croy, R. R. D. (1986). Localization of mRNA for pea legumin: *in situ* hybridization using a biotinylated cDNA probe. *Protoplasma*, **130**, 57–67.

Harris, N., Grindley, H. & Mulcrone, J. (1989). Correlated *in situ* hybridisation and immunocytochemical studies of legumin storage protein deposition in pea (*Pisum sativum* L.). *Cell Biology International Reports*, **13**, 23–35.

Hauxwell, A. J., Corke, F. M. K., Hewdley, C. L. & Wang, T. L. (1989). *In situ* hybridisation for studying cellular development in *Piscum sativum* embryos. *Society for Experimental Biology Edinburgh Meeting Abstracts*, p.6.34.

Hepher, A., Boulter, M. E., Harris, N. & Nelson, R. S. (1988). Development of a superficial meristem during somatic embryogenesis from immature cotyledons of soybean. *Annals of Botany*, **62**, 513–20.

Higgins, T. J. V. (1984). Synthesis and regulation of major storage proteins in seeds. *Annual Review of Plant Physiology*, **35**, 191–221.

Hutchinson, N. (1984). Hybridisation histochemistry: *in situ* hybridisation at the electron microscope level. In: *Immunolabelling for Electron Microscopy*, J. M. Polak & I. M. Varndell, eds, pp. 341–51. Elsevier, Amsterdam.

Kamalay, J. C. & Goldberg, R. B. (1980). Regulation of structural gene expression in tobacco. *Cell*, **19**, 935–46.

Marinos, N. (1970). Embryogenesis of the pea (*Pisum sativum*) I. The cytological environment of the developing embryo. *Protoplasma*, **70**, 261–79.

McFadden, G. I. (1989). *In situ* hybridization in plants: from macroscopic to ultrastructural resolution. *Cell Biology International Reports*, **13**, 3–21.

Meinke, D. W. (1986). Embryo-lethal mutants and the study of plant embryo development. *Oxford Surveys of Plant Molecular and Cell Biology*, **3**, 122–65.

Nelson, T. & Langdale, J. A. (1989). Patterns of leaf development in C4 plants. *Plant Cell*, **1**, 3–13.

Raghaven, V. (1981). Distribution of Poly (A)- containing RNA during normal pollen development and during induced pollen embryogenesis in *Hyoscyamus niger*. *Journal of Cell Biology*, **89**, 593–606.

Raghaven, V. (1986). *Experimental Embryogenesis in Angiosperms*, Cambridge: Cambridge University Press.

Reeve, R. M. (1948). Late embryogeny and histogenesis in *Pisum*. *American Journal of Botany*, **35**, 591–602.

Shirsat, A., Wilford, N., Croy, R. R. D. & Boulter, D. (1989*a*). Sequences responsible for the tissue specific promoter activity of sea legumin gene in tobacco. *Molecular and General Genetics*, **215**, 326–31.

Shirsat, A., Wilford, N. & Boulter, D. (1989*b*). Gene copy number and levels of expression in transgenic plants of a seed specific gene. *Plant Science*, **61**, 75–80.

K. G. Jones, S. J. Crossley and H. G. Dickinson

Investigation of gene expression during plant gametogenesis by *in situ* hybridisation

Introduction

The study of the genes responsible for pollen and ovule development has long been frustrated by the fact that the cells concerned are embedded deep in the sporophytic tissue. The situation is further complicated by their investment by actively transcribing nurse tissues – the tapetum and nucellus. *In situ* hybridisation thus represents the only approach by which gene expression may be studied in these systems. However, the resolution offered by isotopic methods is insufficient to distinguish between members of the microscope tetrad, or individual cells of the developing embryosac. For this reason non-isotopic techniques have been investigated and a protocol devised for anthers. Enzymic detection systems can be used, but high endogenous levels of phosphatase makes blocking and other pretreatments often essential. Further, the methods normally employed to fix and embed material for *in situ* work do not give good results with reproductive cells, principally because of their small size and fragility. The use of resins as embedding media has therefore been explored.

The problems posed by plant reproductive systems

The cells involved in male and female gametogenesis in higher plants are unusual in a number of ways. Firstly, they are comparatively few in number and are surrounded by other tissues carrying out radically different functions. Secondly, they undergo a complex programme of differentiation and division leading to the formation of a number of highly differentiated daughter cells. Thirdly, this development takes place comparatively rapidly, with the cells passing from stage to stage in hours, rather than days as is normally the case for plants. Summarising events within the anther (see Fig. 1), the archesporial tissue first differentiates into a mass of premeiotic pollen mother cells (PMCs) invested, generally, by a single layer of tapetum. Once the PMCs have completed meiosis, the individual haploid microspores are released from the tetrad into the thecal fluid. It is at this point that the pollen

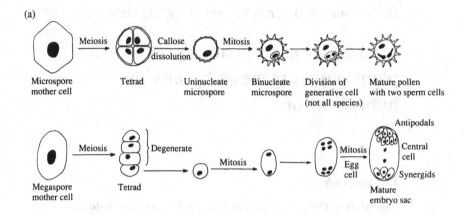

Figure 1. Schematic representation of male (a) and female (b) gametogenesis of higher plants (see text for details).

wall completes its development, and the cytoplasm undergoes the asymmetric first pollen mitosis, forming generative and vegetative cells. A period of intense accumulation of reserves follows, which may take the form of lipid, carbohydrate or protein. The grain is then coated with a tapetally derived pollenkitt or tryphine, dehydrated and released. In some species, the generative cell which, following its formation, moves to a central position within the vegetative cell, may divide again prior to release to form two sperms. Equally spectacular differentiation also takes place in the tapetal layer, where DNA levels may rise to 6 c (Moss & Heslop-Harrison, 1967; Liu, Jones & Dickinson, 1987). The tapetal cells themselves also divide, and sometimes become binucleate. Later in development, generally at or about the tetrad stage, the tapetum in some taxa becomes invasive and permeates the locular space, investing the developing grains. As mentioned above, the tapetum finally contributes to the pollen grain coating covering the grains prior to their discharge.

In the ovule, one single hyperdermal cell develops into the megaspore mother cell, undergoes meiosis and forms a linear tetrad of haploid megaspores (see Fig. 1). Frequently, three of these degenerate, and that remaining – usually at the micropylar end of the ovule – differentiates into the embryosac. This is achieved by a sequence of divisions producing a number of haploid nuclei which are translocated to specific sites within the developing gametophyte. Thus are formed the two synergids, the egg cell itself, the paired haploid nuclei of the central cell – which later forms the triploid endosperm on fertilisation – and a variable number of antipodal cells which are believed to play a nutritional role. As in the case for male development, the developing female cells are invested by a monolayer of secretory cells, termed the nucellus. While nuclear development is not as dramatic as that of the tapetum, the programme of gene expression in this tissue must differ greatly from that of the neighbouring female cells.

These complex patterns of development taking place in cells buried deep in other tissues has always resulted in their investigation being the province of cytologists, cytochemists and most recently, electron microscopists. Thus, any new method devised to study gene expression in these cells must include the ability to examine the cells in question, and determine precisely their developmental stage.

Apart from the major changes in development described above, the first microscopic examinations of plant reproductive cells suggested that a number of fundamental changes may be taking place. For example, Guillermond (1924) and Wagner (1927) both commented on a cycle of differentiation within the 'chondriome' – a term then used for the general organellar content of the cell. Wagner interpreted these changes as representing the elimination of organelle line during gametogenesis, whereas Guillermond drew the opposite conclusion. Py (1932) and others described dramatic changes taking place in the organellar population of female cells, whilst Painter (1943) and Sauter & Marquardt (1967) reported unusual changes in the 'stainability' of sex cell cytoplasm. Most recently, a combination of electron microscopy and accurate cytochemistry has revealed that organelle line persists throughout meiosis and mega/microporogenesis in higher plants (Marumaya, 1968; Dickinson & Heslop-Harrison, 1970b) and that the change in cytoplasmic stainability reported results from a comprehensive expunging of RNA from the cytoplasm of these cells (McKenzie, Heslop-Harrison & Dickinson, 1967; Dickinson & Heslop-Harrison, 1970a; Williams, Heslop-Harrison & Dickinson, 1973). Meiotic biochemistry has been studied over the past two decades by Stern and co-workers, and a small proportion of the genome – late replicating in zygotene and pachytene – has been proposed to play a central role in pairing and recombination (Stern & Hotta, 1987). Female cells are completely inaccessible to modern biochemical methods and, for this reason, even less is known of their development. However, most of the cytoplasmic changes described for male cells also seem to occur in the female (Dickinson & Potter, 1978) and it is assumed that the meiotic biochemistry is identical.

In view of the above, *in situ* hybridisation thus emerges as the only technique whereby gene expression may be studied in plant reproductive cells. However, it is equally clear that the use of highly radioactive probes on fresh or wax embedded tissue, such as has been so helpful in the study of *Drosophila* development (Hafen & Levine, 1986), is not applicable to these plant systems. For example, in many cases it is necessary to distinguish between expression in adjacent cells, and thus the high energy emission of ^{35}S or ^{32}P provides a confused picture. Equally, any method employed must involve the accurate preservation of material, so that individual cell lines may be identified and their stage of development determined. Finally, in view of the rapidity of plant sex cell differentiation it may sometimes be necessary to quantify levels of expression in particular cells at specific times.

We describe here some new methods for the fixing and embedding of plant material for *in situ* hybridisation which give excellent tissue preservation with a minimal loss of signal. We have also evaluated a range of biotin-based non-isotopic detection and revealing systems and draw conclusions as to those which are most suitable for use in these circumstances. Quantitation of *in situ* hybridisation using a synthetic probe is also described, and, finally, we report another *in situ* method which is proving of considerable application in the study of plant reproductive systems.

The development of non-isotopic high resolution *in situ* systems for plant reproductive cells

Although isotopic detection systems are ideal for quantification and the detection of high abundance poly A + RNA, the exposure times involved in the detection of low abundance message are time consuming, and permit the accumulation of background. In addition, if a label other than ^{3}H is used, extensive 'cross-fire' occurs, with a consequent decrease in resolution of the system. We have investigated the use of biotin-based non-isotopic detection systems using a range of visualisation methods. In comon with the majority of groups working in this area, we appreciate that the most efficient and sensitive probes for *in situ* systems are sense/antisense RNA (see Chapters 3, 4, this volume) but, for convenience in this preliminary work, we have used as a probe a 0.9 kb EcoRI/Bam HI DNA fragment derived from a genomic clone encoding the ribosomal genes of wheat (Gerlack & Bedbrook, 1979). The probe was normally labelled by nick translation with biotinylated dUTP, but we have also experimented with photobiotinylation and found it also to be satisfactory.

A high resolution detection system is only useful when cellular detail is well preserved. We have examined the effect of a range of fixatives on the male meiocytes of *Lilium*, and found that, whilst depolymerised paraformaldehyde probably preserves more signal, a mixture of low concentrations of glutaric dialdehyde with about 4% paraformaldehyde represents the best compromise between tissue and signal preservation. The embedding medium used may also dramatically affect the preservation of the tissue. Figure 2(a) and (b) shows an anther of *Lilium* embedded in paraplast and stained with acridine orange. Whilst RNA is clearly well preserved, consider-

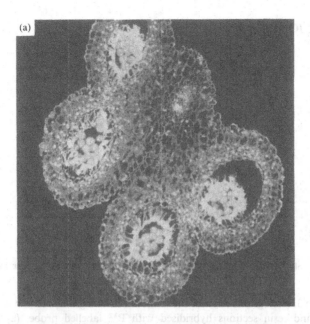

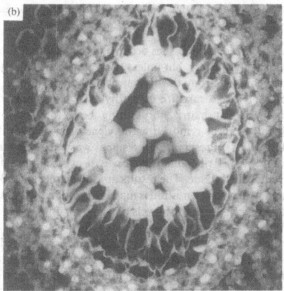

Figure 2. 2(a) and (b). Transverse section of *Lilium* anther at tetrad stained with acridine orange and viewed under UV illumination. (a) Whole anther, demonstrating retention of nucleic acids in fixed and paraplast-embedded material. Note the shrinkage of locular material away from anther wall layers. (b) Higher magnification of Figure 2(a) showing a single locule. Flurorescence due to RNA is clearly visible in the cytoplasm of tapetal cells, and fluorescence due to DNA is seen in the nuclei. The two types of fluorescence can be readily differentiated on the basis of colour (orange for RNA, yellow for DNA).

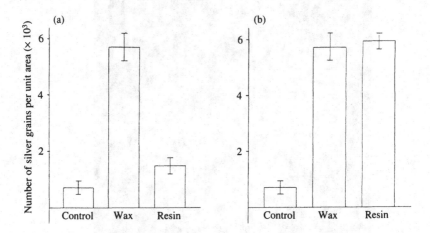

Figure 3. Histograms to show the number of silver grains per unit area on wax and resin sections hybridised with P^{32} labelled probe. (a) Non-compensated. (b) Compensated for differences in section thickness.

able shrinkage has occurred in the tissues occupying the loculus. Embedding in resin provides far better results, but unless the resin is particularly amorphous, or can be removed from the tissue, access of the probe will be inhibited. We routinely use a mixture of methyl and butyl methacrylates which provide an excellent matrix for cutting both for electron light microscopy, and which can be removed with xylene prior to hybridisation (Parks & Spradling, 1987). Preservation of signal was clearly a matter of concern, so the binding of ^{32}P ribosomal probe was measured by grain counts on wax and resin embedded tissue. The results are shown in Figure 3(a) and (b) and indicate that, when compensation is made for section thickness, there is no significant difference between the levels of signal retained following the two treatments.

A standard protocol for blocking and hybridisation was adopted in this work, and is set out in Appendix 1. Attempts were made to optimise each of the stages, but since it was appreciated that eventually RNA probes would be employed, it was considered unnecessary to devote much effort to maximising hybridisation of the DNA probe used.

A number of methods for detecting the presence of bound probe were investigated. Since the probe contained biotin, the detection systems available were based on the binding either of avidin/streptavidin, or antibiotin. Both these approaches were tried, together with enzymic and gold-based detection systems (see Appendix 2). A number of important factors emerged from this work; firstly, peroxidase did not work well in our hands for,

Gene expression during plant gametogenesis

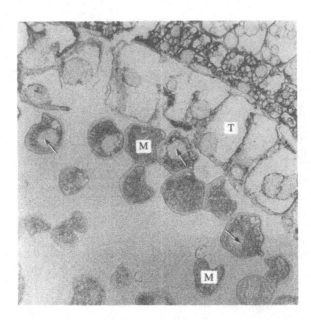

Figure 4. Bright field micrograph of a transverse section (2 μm) of an anther locule of *Lilium* probed with biotin-labelled DNA encoding rRNA, and visualised using a streptavidin–biotin–alkaline phosphatase detection system. Labelled probe is concentrated specifically over the cytoplasm of both tapetal (T) and meiocyte (M) cells, whilst nuclei (arrows) are unlabelled.

although it was capable of extreme sensitivity, it was unreliable and we had little confidence that the level of reaction was at all related to the signal present. Secondly, methods involving streptavidin–gold with silver enhancement, while sensitive, also were so unstoichometric that they were of little use. Most of our work has been carried out using alkaline phosphatase, which produces a reliable, strong reaction which we believe to be related to the amount of signal present (see Fig. 4). Extreme care must, however, be exercised when using alkaline phosphatase detection systems on reproductive cells because the enzyme itself is a well-known 'marker' for germinal tissue in animals, and is certainly a feature of male and female meiocytes in plants. The native enzyme was certainly a considerable problem in wax-embedded tissue and resulted in other methods having to be adopted. However, following the processing necessary for resin embedding, the bulk of the native alkaline phosphatase activity is lost from the cells, and thus it may be confidently used as part of a detection system.

Most recently, antibiotin protein A-gold has been used as an *in situ*

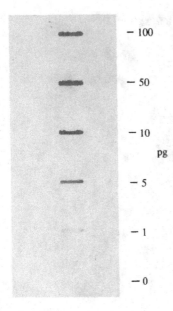

Figure 5. Immunogold detection of λ DNA by slot-blot hybridisation with biotin-labelled probe. Varying amounts (pg) of unlabelled λ DNA were immobilised on nitrocellulose and hybridised with biotin-labelled λ. Hybrids were detected using anti-biotin antibody and protein A-gold, with silver enhancement.

detection system at both the light and electron microscope levels. It has proved highly reliable, and can readily detect levels of target down to 1 pg in filter hybridisations (see Fig. 5). Further details of the sensitivity of this probe for work with plant tissues will be reported elsewhere (K. G. Jones and H. G. Dickinson, in preparation).

Quantification of *in situ* hybridisation

In circumstances where high levels of a specific message are known to be present, and when a reliable probe is available, there is no reason why the degree of hybridisation cannot be measured *in situ*. Not all methods of detection can be utilised for this purpose, since a direct relationship must exist between the probe hybridised in the tissue, and the signal observed or measured. For this reason multistage amplification systems of detection, and methods involving enhancement with silver cannot generally be used owing to their lack of stoichiometry. Isotopically labelled probes are best used in these circumstances and we have adopted this approach with our work.

However, as has already been pointed out, ^{35}S and ^{32}P labelled probes activate silver grains over such a large area that it is often difficult to determine in which cell the target polynucleotide resides. Using ^{3}H probes, accurate localisation and quantification can be achieved, albeit at the price of an extended exposure time. There is no doubt that background increases with exposure period, but the use of a probe with high specific activity and the thinnest possible film of emulsion will normally ensure acceptable background levels.

Our earliest work in this area involved the simplest type of *in situ* hybridisation possible, the detection of all cellular polyadenylated RNA using tritiated synthetic polyuradylic acid. This work is described in detail in two publications (Porter, Parry & Dickinson, 1983; Dickinson & Wilson, 1985) but the principal features merit repetition here. Plant reproductive cells have been known for some time to exhibit unusual RNA metabolism (McKenzie, Heslop-Harrison & Dickinson, 1967; Dickinson & Heslop-Harrison, 1970a; Williams, Heslop-Harrison & Dickinson, 1973; Porter, Bird & Dickinson, 1982) and that, in addition to the near elimination of ribosomal RNA, synthesis of poly A+ RNA also rapidly decreases. Measurement of synthesis gives no indication of total cellular pools of poly A+ RNA, and since the investing tapetal cells clearly contain high levels of ribosomal and poly A+ RNA, the only way open to us to estimate total poly A+ RNA was to employ an *in situ* approach. At the same time, we were also interested in the nature of nucleolus-like bodies which appear in the male and female meiocytes immediately following meiosis (see Fig. 6). These cytoplasmic nucleoloids seem to be restricted to reproductive tissue and disintegrate during early micro- and megaspore development. Their chemistry is obscure for although they stain well with the silver reagent used for nucleoli, their response to acridine orange and other RNA specific stains is often equivocal. Nevertheless, the possibility remains that they constitute some form of 'informasome', comprising both ribosomes and poly A+ RNA which could determine the early stages of microspore development.

Meiocytes of *Lilium henryii* were fixed in a mixture of glutaraldehyde and paraformaldehyde and embedded in water soluble methacrylate resin. Following sectioning at either 0.5 or 2 μm, the resin was removed and the material hybridised with commercially obtained [^{3}H] poly-U. After appropriate washing stages the material was dried and covered with a monolayer of Ilford L4 emulsion, using the loop method. For electron microscopy, gold refracting sections were cut, placed on grids, and hybridised. These were then coated with Ilford G5 emulsion using the same procedure. Exposure times of between 2–4 weeks (LM) and 3–5 months (EM) were found necessary to provide acceptable levels of signals. Full details of this procedure are given in Porter *et al.* (1983).

Following development, the slides and grids were examined and silver

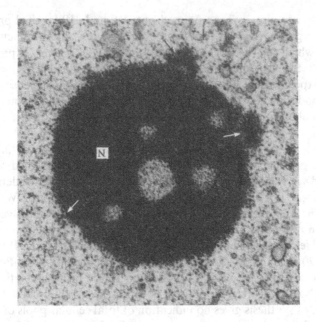

Figure 6. Cytoplasmic nucleoloid (N) in male meiocyte of *Lilium* TEM. Note the apparent production of ribosomes (arrows) from the nucleoloid surface.

grains counted over the cells or structures of interest. Using conventional methods for assessing the results from autoradiography, the levels of binding of the probe to meiocytes and to the cytoplasmic nucleoloids were determined. Interestingly, the total cellular pool of poly A+ RNA was found to decrease in a similar fashion to ribosomal RNA, and be restored to normal levels at approximately the same time. Careful inspection of the cytoplasmic nucleoloids revealed that they contained no poly A+ RNA. Indeed, our subsequent researches (Sato, Willson & Dickinson, 1989) indicates that these structures contain little or no RNA whatsoever, but are accumulations of meiosis-specific proteins, previously associated with the chromosomes. These inferences were drawn following the use of RNAase gold as a probe, which provides a very accurate measurement of RNA levels within subcellular structures (Sato et al., 1989).

Although this early *in situ* work was simple in the extreme, the fact that it could be checked against poly A+ RNA synthesis, and indeed against the total levels of RNA within the cell, provided some indication as to its accuracy. Further, although the results with regard to the cytoplasmic nucleoloids were in a way disappointing, the method certainly proved capable of detecting *in situ* hybridisation within a cytoplasmic structure of some 0.25 μm in diameter.

In situ nick translation in plant reproductive cells

There is currently considerable controversy concerning the transmission of organelles through the male gamete in flowering plants. It has been known for some while that some plants transmit plastids in this fashion whilst others do not, but the mechanism by which organelles are eliminated from the gametes of 'non-transmitting' species is unclear. Recent evidence suggests that a nuclease active late in pollen maturation degrades organellar DNA in the 'non-transmitters', and this view has been supported by the comprehensive light microscopic investigation of Miyamura, Kuroiwa & Nagata, 1987. This work, in which DAPI was used to detect DNA, demonstrated eradication of organellar genomes to occur in the generative cells of a number of angiosperms. However, we know from electron microscopic studies that mitochondria, at least, are present throughout the life of the generative and sperm cells. There is therefore the possibility that 'during this period' the levels of organellar DNA decrease, to such an extent that they are undetectable using DAPI or similar fluorochromes, but yet are not completely eliminated. To test this hypothesis, a method was devised for the detection of low levels of DNA in sections cut for the electron microscope.

There are two approaches to this problem, one being the use of a DNA antibody, and the other *in situ* nick translation on the section itself. While antibody methods are certainly very sensitive, the 'anti-DNA' antibodies with the highest activity are those raised to DNA/RNA hybrids, and their use would be inappropriate in this investigation. Similarly, antibodies would also tend to pick up short chain polynucleotides – perhaps formed from the degradation of DNA – and thus would wrongly give the impression of intact organellar genomes.

For this reason we adapted a conventional nick translation system to work on material embedded in LR gold resin, and cut to a thickness of 150 nm. Biotinylated dUTP was supplied during the reaction, and the newly-formed polynucleotide was detected using antibiotin protein A-gold. Full details of the procedure employed are set out in Appendix 3. A number of controls were carried out, all of which displayed no specific staining either over mitochondria or the nucleus. Figure 7(a) and (b) shows a typical preparation, together with a control, with gold particles localised apparently over the mitochondrial genome. The background was, generally, very low and, as might be expected, considerable quantities of probe bound to the nucleus. Work is currently now in hand to determine the relative sensitivities of immunological and *in situ* nick translation methods for the detection of DNA in sections.

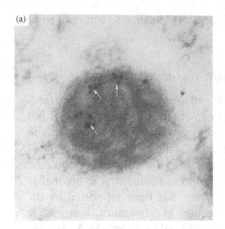 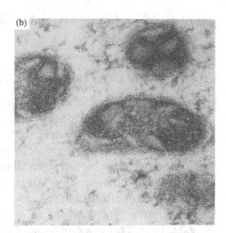

Figure 7. Biotin labelling of organellar DNA by *in situ* nick translation. (a) Electron micrograph of meiocyte mitochondrion in ultrathin section labelled by nick translation with biotin dUTP and detected with immunogold. Gold label (arrows) is present within the organelle, whilst background cytoplasmic staining is negligible. (b) Control section of 7(a) incubated with DNase prior to labelling and detection.

Acknowledgements

The authors gratefully acknowledge financial support from the University of Reading Research Endowment Fund.

References

Dickinson, H. G. & Heslop-Harrison, J. (1970a). The ribosome cycle, nucleoli and cytoplasmic nucleoloids in the meiocytes of *Lilium*. *Protoplasma*, **69**, 187–200.

Dickinson, H. G. & Heslop-Harrison, J. (1970b). The behaviour of plastids during meiosis in the microsporocytes of *Lilium longiflorum* Thunb. *Cytobios*, **6**, 103–118.

Dickinson, H. G. and Potter, U. (1978). Cytoplasmic changes accompanying the female meiosis in *Lilium longiflorum* Thunb. *Journal of Cell Science*, **29**, 147–9.

Dickinson, H. G. & Willson, C. (1985). Behaviour of nucleoli and cytoplasmic nucleoloids during meiotic divisions in *Lilium henryi*. *Cytobios*, **43**, 349–65.

Gerlach, W. L. & Bedbrook, J. R. (1979). Cloning and characterisation of ribosomal DNA from wheat and barley. *Nucleic Acids Research*, **7**, 1869–85.

Guilllermond, A. (1924). Recherches sur l'évolution du chondrisome pendant le development due sac embryonaire et des cellules mères des graines de pollen dans les Liliacees et la signification des formations ergastoplasmiques. *Annales des Sciences Naturelle, Botanique*, **6**, 1–52.

Hafen, E. & Levine, M. (1986). *In situ* hybridisation to RNA. In *Drosophila – A Practical Approach* (ed. D. B. Roberts), pp. 139–57, London: IRL Press.

Liu, X. C., Jones, K. G. & Dickinson, H. G. (1987). DNA synthesis and cytoplasmic differentiation in tapetal cells of normal and cytoplasmically male sterile lines of *Petunia hybrida*. *Theoretical and Applied Genetics*, **74**, 846–51.

Marumaya, L. (1968). Electron microscopic observation of plastids and mitochondria during pollen development in *Tradescantia paludosa*. *Cytologia*, **37**, 482–97.

Mackenzie, A., Heslop-Harrison, J. & Dickinson, H. G. (1967). Elimination of ribosomes during meiotic prophase. *Nature London*, **215**, 997–9.

Myamura, S., Kuroiwa, T. & Nagata, T. (1987). Disappearance of plastid and mitochondrial nucleoloids during the formation of generative cells of higher plants revealed by fluorescence microscopy. *Protoplasma*, **141**, 149–59.

Moss, G. I. & Heslop-Harrison, J. (1967). A cytochemical study of DNA, RNA and protein in the developing maize anther. II. Observations. *Annals of Botany*, **31**, 555–72.

Painter, T. S. (1943). Cell growth and nucleic acids in the pollen of *Rhoeo discolor*. *Botanical Gazette*, **105**, 58–68.

Parks, S. and Spradling, A. (1987). Spatially regulated expression of charion genes during *Drosophila* oogenesis. *Genes and Development*, **1**, 497–509.

Porter, E. K., Bird, J. & Dickinson, H. G. (1982). Nucleic acid synthesis in microsporocytes of *Lilium* cv. cinnabar: events in the nucleus. *Journal of Cell Science*, **57**, 229–46.

Porter, E. K., Parry, D. & Dickinson, H. G. (1983). Changes in poly(A)$^+$ RNA during male meiosis in *Lilium*. *Journal of Cell Science*, **62**, 177–86.

Py, G. (1932). Recherches cytologiques sur l'assis nouriciere des microspores et les microspores des plantes vasculaires. *Revue generale de Botanique*, **44**, 316–413, 450–62.

Sato, S., Willson, C. & Dickinson, H. G. (1989). The RNA content of the nucleolus and nucleolus-like inclusions in the anther of *Lilium* using an improved RNase-gold labelling method, in press.

Sauter, J. J. & Marquardt, H. (1967). Die Rolle des Nukleohistons bei der RNS- und Proteinsynthese während der Mikrosporogenese von *Paeonia tenuifolia*. *Zeitschrift für Pflanzenphysiologie*, **58**, 126–37.

Stern, H. & Hotta, Y, (1987). The biochemistry of meiosis. In *Meiosis: Cell Biology* (ed. P. Moens, pp. 303–31. London: Academic Press.

Wagner, N. (1927). Evolution du chondriome pendant la formation des graines de pollen des angiosperms. *Biologia generalis*, **3**, 15–66.

Williams, E., Heslop-Harrison, J. & Dickinson, H. G. (1973). The activity of the nucleolus organising region and the origin of cytoplasmic nucleoloids in meiocytes of *Lilium*. *Protoplasma*, **77**, 79–93.

Appendix 1. *Protocol for* in situ *hybridisation*

Slides	× Acid-washed 3-aminopropyltriethoxysilane-coated
Specimen	5–10 μm paraffin sections 2 μm methacrylate resin sections
Dry	55 °C overnight
Hydration	2 × 10 min xylene, hydrate in graded ethanol
Hydrolysis	10 min 0.2 M HCl
Wash	2 × 5 min 1X SSPE
Proteolysis	20 min 100 μg/ml autodigested pronase, 50 mM Tris Cl, pH 7.2, 5 mM EDTA 37 °C
Wash	2 × 5 min 1X SSPE
Fixation	20 min 4% paraformaldehyde, 1 × SSPE
Wash	2 × 5 min H_2O
Acetylation	10 min 0.25% acetic anhydride in 0.1 M triethanolamine, pH 8.0
Wash	2 × 5 min H_2O
Dehydration	dehydrate in graded ethanol
Add probe	2–4 μg/ml in hybridisation buffer (a), 30 μl per slide
Hybridise	24–48 h at 37 °C
Wash	variable, according to desired stringency
Blocking	30 min TBS (b). 3% BSA, 0.5% Tween-20, 37 °C
Detection	(a) Hybridisation buffer: 2X SSC (pH 7), 50% formamide, 10% dextran sulphate, 400 μg/ml salmon sperm DNA, 2X Denhardts solution. (b) TBS: 0.1 M Tris Cl, pH 7.5, 0.1 M NaCl, 5 mM $MgCl_2$

Appendix 2. *Detection systems for biotin-labelled probes*

Detection		Visualisation
Primary conjugate	Secondary conjugate	
Streptavidin	Biotin–Alkaline phosphatase	NBT + BCIP
Avidin–biotin–peroxidase complex	—	DAB + Nickel
Avidin–biotin–peroxidase complex	—	4–chloro, 1-naphthol
Anti-biotin (goat)	Protein A-gold (5 nm)	Silver enhancement
Streptavidin–gold (20 nm)	—	Silver enhancement

Appendix 3. In situ *nick translation on EM sections*

Mount ultrathin EM sections on formvar-coated nickel grids.
Place grids on 20 μl nick translation reaction mixture containing biotinylated dUTP. Incubate 15 °C for 20 min.
Dip grids in 5 mM EDTA, pH 8.
Rinse in droplets of 1 × SSC, 0.1% SDS for 5 × 5 min.
Dip grids in 20 mM Tris, 0.5 M NaCl, 0.1% BSA pH 8.2 (TBS buffer).
Block grids with 10% BSA in TBS, 10 min RT.
Incubate in goat antibiotin (1:50 dilution in TBS) for 2 h RT.
Rinse with TBS from a wash bottle.
Incubate in Protein A conjugated to colloidal gold (5 nm) for 1 h RT.
Wash in TBS.
Wash in distilled water.
Fix the DNA/antibody/conjugate complex in 1% GDS for 10 min RT.
Wash in distilled water.
Stain with uranyl acetate and lead citrate.

John D. West

Sexing the human conceptus by *in situ* hybridisation

Introduction

DNA–DNA *in situ* hybridisation is widely used by geneticists as a tool for mapping genes on chromosomes. More recently, the development of chromosome-specific probes has enabled *in situ* hybridisation to be used to detect the presence of specific chromosomes in interphase nuclei and to determine the number of copies of chromosomes present. This has been applied to clinical prenatal diagnosis of human genetic diseases that involve numerical chromosome aberrations and in cases where only boys are at risk for X-linked disorders. This approach has also been useful as an experimental tool for detecting polyploid interphase nuclei. My own work in this field has been confined to *in situ* hybridisation with a Y chromosome DNA probe to sex the human fetus and pre-embryo. These studies illustrate the possibilities of using DNA–DNA *in situ* hybridisation both for prenatal or preimplantation diagnosis and for experimental investigations of ploidy.

Prenatal diagnosis

At present, samples of cells for prenatal diagnosis may be obtained by amniocentesis, chorionic villus sampling or fetal blood sampling. In the future it may also be possible to sample cells of a preimplantation stage embryo (or pre-embryo) for genetic tests before the pre-embryo is returned to the uterus to continue development. This possibility of preimplantation diagnosis is discussed in a later section.

Amniocentesis involves the aspiration of 10–20 ml of amniotic fluid at 16–20 weeks gestation. At 18–19 weeks, 20 ml of amniotic fluid contains about 1 million cells (Gosden & Gosden, 1985) of heterogeneous origin and yields approximately 7 μg of DNA (Old, 1986). (The theoretical maximum yield for a million cells, each containing 12 picograms of DNA, is 12 μg.) Chorionic villus biopsy is usually performed at 8–12 weeks but this is now being extended to second trimester placental biopsy. This typically yields 5–50 mg of tissue and this has to be cleaned of contaminating maternal cells. In some cases fetal blood is sampled (Rodeck & Campbell, 1978; Nicolaides *et al.*,

1986) and a typical sample will comprise 0.2 to 0.7 ml of blood taken between 17 and 40 weeks. (Between 15 and 21 weeks, fetal blood contains about 2×10^9 white cells per litre and a similar number of nucleated red cells (Millar et al., 1985). By 32–34 weeks the proportion of nucleated red cells has fallen to 0–50 per 100 white cells (Nicolaides et al., 1986).)

These three types of sample can be used for a wide range of cytogenetic, biochemical or molecular tests. Cytogenetic tests require high quality preparations of metaphase cells and biochemical and standard molecular tests require sufficient tissue for a reliable assay or DNA extraction. Amniotic fluid cells may be cultured for two to three weeks to increase the yield. The sensitivity of many molecular tests can be dramatically increased by amplifying the target DNA with the polymerase chain reaction (Mullis & Faloona, 1987; Saiki et al., 1988). DNA–DNA in situ hybridisation can also be used as a genetic test for several types of genetic disorder. This approach requires only a small number of interphase nuclei to determine the sex of a fetus or test for the presence of a suspected trisomy or other numerical chromosome anomaly.

Some sex-linked disorders cannot yet be diagnosed by specific genetic tests so that identification and termination of male fetuses is all that can be offered to many patients who are at risk. Early identification of a male fetus is of obvious benefit to these patients. If a male fetus, at risk for a genetic disorder that can be diagnosed by a specific test, is recognised early in pregnancy, this allows time for a further sample and test (e.g. fetal blood sample to diagnose haemophilia A). Early identification of a female fetus benefits all couples who risk having sons with X-linked genetic disorders (even when the disorder can be identified by a specific genetic test), because they are immediately reassured that they will not have an affected son.

Sexing by DNA–DNA *in situ* hybridisation

Sexing by DNA–DNA *in situ* hybridisation requires a Y chromosome-specific DNA probe that detects a target sequence that is sufficiently repeated to give an unambiguous signal to both metaphase and interphase nuclei after hybridisation. There are several such probes available (Table 1) and the choice is determined by a variety of factors that include availability, number of copies of target DNA sequence on the Y chromosome, lack of significant cross-reacting autosomal target sequences and location of the target sequences on the human Y chromosome. In order to determine whether the individual is morphologically a male, probes close to the testis-determining factor gene (TDF) on the short arm (Page et al., 1987) would be most suitable. This would enable us to identify XX males as males if they retained a portion of the Y chromosome that included the target DNA as well

Table 1. *DNA probes suitable for identification of specific chromosomes in interphase nuclei by in situ hybridisation*

Chromosome	Probe	Label[a]	References
Amniotic fluid cells, chorionic villus cells, lymphocytes, etc			
9	pHuR98	A,B	Pinkel *et al.* (1986*a*), Trask *et al.* (1988)
17	p17H8	B	Trask *et al.* (1988)
18	L1.84	A,B,T	Cremer *et al.* (1986)
18	pBRHS13	T	Vorsanova *et al.* (1986)
21	—	B	Julien *et al.* (1986)
X	pXBR-1	T	Rappold *et al.* (1984)
X	pSV2X-5	B	Fantes *et al.* (1988)[b]
Y	CY1	T	Rappold *et al.* (1984)
Y	pHY2.1	B	Burns *et al.* (1985)
Y	pHY2.1	T	West *et al.* (1989*a*)
Y	pY3.4	T	Lau *et al.* (1985)
Y	pY3.4A	A	Pinkel *et al.* (1986*a*), Trask *et al.* (1988)
Y	pY431	B	Pinkel *et al.* (1986*b*)
Y	pY431a	B	Guyot *et al.* (1988)
Y	Y190	B	Kozma & Adinolfi (1988)
Y	pDp105	T	Disteche *et al.* (1986)[b]
Pre-embryos			
Y	pHY2.1	T,B	West *et al.* (1987,1988)
Y	p102d	T	Jones *et al.* (1987)
Y	pHY2.1	B	Handyside *et al.* (1989), Penketh *et al.* (1989)

[a] A, AAF(2-acetylaminofluorine) modification of guanosine in DNA; B, biotin incorporation; T, tritium incorporation.
[b] Not used with interphase nuclei.

as the TDF gene. However, for prenatal diagnosis we are usually more concerned with whether an individual has inherited an X chromosome from the father as well as from the mother because, in most cases at risk for X-linked diseases, the abnormal gene is passed from the mother's X chromosome and only male offspring that are hemizygous for the defective gene are affected. In this case it is more useful to detect the presence of two X chromosomes and/or the absence of a Y chromosome. Although there are some X chromosome probes (Table 1), in practice it is probably easier to sex interphase nuclei by the presence or absence of a single Y hybridisation site than to distinguish between 1 or 2 X chromosome hybridisation sites. Although, double labelling with X and Y chromosomal DNA probes is possible (for example one labelled with tritium and the other with a non-radioactive label such as biotin), at present fetal sexing by DNA-DNA *in situ* hybridisation is almost always done with a Y chromosome probe. Ideally the Y probe should detect reiterated DNA sequences close to the centromere of the Y chromosome so that even Y chromosomes with small, long arms are detected. However, prior hybridisation to parental lymphocyte samples will forewarn of any inherited polymorphisms such as small Y chromosomes or Y/autosome translocations (see below). The incidence of such polymorphisms arising *de novo* is low.

For our studies we have used Y probe pHY2.1 (Cooke *et al.*, 1982) or the identical commercially available probe, Amprobe RPN 1305X (Amersham). This probe recognises a 2.12 kb sequence that is repeated 2000 times on the long arm of the Y chromosome. Most of these sequences are located on the distal fluorescent region of the Y chromosome but hybridisation is still detectable to Y chromosomes that lack the fluorescent region (Gosden *et al.*, 1984). An additional 100 copies of a related 2.00 kb sequence are located elsewhere in the genome but produce a much weaker signal after *in situ* hybridisation than those on the Y chromosome. Most of our work has been done with tritiated probe on various cell types that have been treated in hypotonic and fixative and then spread on microscope slides. In some cases biotinylated Y probe was used as discussed on pages 223-6. The hybridisation methods are described in the Protocols and other details are given elsewhere (West *et al.*, 1987, 1988, 1989*a*, *b*).

Sexing the fetus

After *in situ* hybridisation to the Y probe, the presence of the Y chromosome can be detected as a large hybridisation site (hybridisation body or Y body) in both interphase and metaphase cells. This appears as a cluster of silver grains after hybridisation with tritiated probe and autoradiography (Fig. 1) or, if the probe is non-radioactively labelled, the hybridisation bodies are eventually visualised either as coloured precipitates or as fluorescent sites. For routine analysis we scored 100 interphase nuclei and classified them according to the number of hybridisation bodies present (0, 1?, 1, or >1 per nucleus; a doubtful or weak hybridisation body was classified as 1?).

Figure 2 shows the frequency of labelled nuclei from male and female control preparations. The proportion of unlabelled nuclei among the XY controls varies between experiments (typically 60-100%) but the proportion of false positive nuclei among the XX controls is rarely more than a few per cent. This provides excellent discrimination between preparations from males and females. Further confirmation that the clusters of grains identify the Y chromosome is shown in Figure 3. This demonstrates that the Y probe labels the region of the Y chromosome that fluoresces after exposure to spermidine *bis*-acridine.

The results for sexing unknown samples of adult lymphocytes, fetal lymphocytes, chorionic villus cells and amniotic fluid cells are summarised in Figure 4 and examples are illustrated in Figure 1. A total of 82 samples were scored and independent cytogenetic sexing was done on 78 of these samples. Both methods were in complete agreement (West *et al.*, 1989*a*). Figure 4 shows that there is a clear distinction between male and female samples when 100 interphase nuclei are scored. Compared with the other cell types, the amniotic fluid cells from the male conceptuses had a higher proportion of nuclei with no hybridisation body and a higher frequency of nuclei with more than one hybridisation body (Figs 4 and 5). This greater heterogeneity reflects the heterogeneous origin of amniotic fluid cells (Gosden, 1983). Nuclear measurements (West *et al.*, 1989*a*) suggest that the amniotic fluid cells with two hybridisation bodies are probably tetraploid nuclei (Table 2).

These results show that DNA-DNA *in situ* hybridisation and analysis of interphase nuclei provides an accurate alternative to cytogenetic analysis of metaphase spreads for fetal sexing. Although this method cannot replace cytogenetics because a full cytogenetic analysis yields a wealth of information in addition to sex determination, *in situ* hybridisation provides an important aid to diagnosis. *In situ* hybridisation can be done on primary samples of tissue and so avoid problems associated with culture (such as overgrowth by maternal cells and culture-induced mosaicism). It may be particularly useful 1) when there are insufficient dividing cells for a reliable cytogenetic analysis, 2) when rapid sex determination is required for a fetus at risk for a sex-linked condition (prolonged cell culture is not required), and 3) to help identify 45,X/46,XY mosaicism and genuine polyploid/diploid mosaicism (genuine

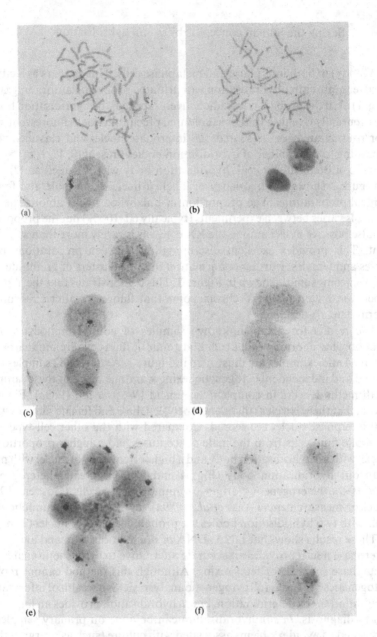

Figure 1. Hybridisation of tritiated Y probe to nuclei from male ((a), (c), (e)) and female ((b), (d), (f)) conceptuses. Samples are from amniotic fluid cells ((a), (b)), chorionic villus cells ((c), (d)) and fetal blood ((e), (f)). The hybridisation bodies, indicating the presence of the Y chromosomes are seen as discrete clusters of silver grains over the interphase nuclei (in (a), (c) and (e)) and over the Y chromosome (in (a)). Reproduced with permission from West et al., 1989a.

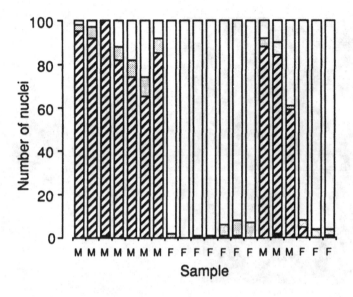

Figure 2. Incidence of nuclei with 0 (□), 1 equivocal (▨), 1 (▨), or >1 (■) Y hybridisation bodies from control slides hybridised to tritiated Y probe. The first 14 samples were male (M) and female (F) lymphocyte cultures and the remaining 6 samples were lymphoblastoid cell lines. Reproduced with permission from West *et al.*, 1988.

tetraploid/diploid mosaicism may be associated with fetal anomalies but similar mosaicism may also arise during cell culture).

When a Y probe is used clinically to sex conceptuses by *in situ* hybridisation to interphase nuclei, it is important to test blood samples from both parents to screen for any abnormalities. For example, interphase nuclei from a female conceptus with a Y/autosome translocation would have a single hybridisation body and thus, on this basis, be scored as a male, unless the parents are screened or cytogenetic sexing is also done. Many interphase nuclei of male Y/autosome translocation carriers have two Y bodies (P. M. Ellis, J. D. West, K. M. West, R. S. Murray and M. C. Coyle unpublished data) as shown in Figure 5. (Other nuclei with two Y hybridisation bodies include polyploid nuclei (see above, those from 47,XYY males (C. M. Gosden, J. D. West and K. M. West unpublished data) and some interphase nuclei which carry a dicentric Y chromosome (E. Grace, P. M. Ellis, J. D. West and M. Keighren, unpublished data).) The frequencies of 46,XY t(Y;14) and 47,XYY nuclei with two hybridisation bodies (44.6% and 49.0%) were lower than the frequencies of control male nuclei with single hybridisation bodies (66.7% and 73.5% respectively).

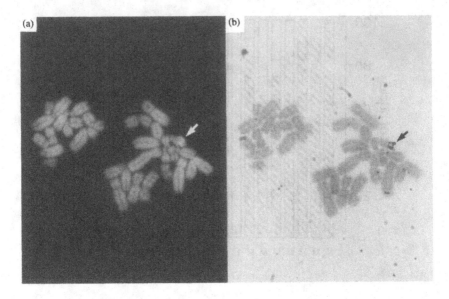

Figure 3. Part of a metaphase spread from a control male lymphocyte culture (a) pretreated with spermidine *bis*-acridine to reveal the fluorescent Y-chromosome then (b) hybridised to the tritiated Y probe to reveal specific hybridisation to the Y chromosome. Reproduced with permission from West *et al.*, 1988.

Sexing the pre-embryo: a possible test for preimplantation diagnosis

Pre-implantation stage human embryos (or 'pre-embryos'; McLaren, 1986) are now routinely cultured to the four to six-cell stage during the course of *in vitro* fertilisation therapy for infertility and later stages (e.g. blastocysts) can be flushed from the reproductive tract and returned for implantation into the uterus (e.g. Buster *et al.*, 1985). A 7-cell human pre-embryo is illustrated in Figure 6. If one or more cells could be safely removed from such a pre-embryo (without compromising the pre-embryo's potential for further development) it would be possible to carry out prenatal diagnosis at the preimplantation stage (McLaren, 1985; West *et al.*, 1987; West, 1989; Penketh & McLaren, 1987; Monk, 1988). It would then be possible to avoid transferring genetically affected pre-embryos to the uterus for further development, thus avoiding the need for therapeutic abortion of affected conceptuses identified at later stages of development. Realisation of this technique requires: 1) the development of a safe method of removing a cell (or

Table 2. *Relative sizes of amniotic fluid cell nuclei with one or two hybridisation bodies (data from West et al. 1989a)*

Sample	Number of Y bodies per nucleus	Number of nuclei scored	Mean nuclear area (μm^2)	Area ratio (2Y:1Y)
A	1	23	719	
A	2	23	1,203	1.67
B	1	22	759	
B	2	22	1,408	1.86
C	1	20	588	
C	2	20	1,223	2.08
D	1	9	508	
D	2	9	1,031	2.03
E	1	9	769	
E	2	9	1,151	1.50

The surface area ratio of two spheres with a volume ratio of 2:1 is 1.58:1. After spreading on a microscope slide, the area of tetraploid nuclei is expected to be between 1.58 and 2 times that of diploid nuclei of the same cell type (see West et al., 1989a).

cells) from human pre-embryos, 2) the use of cryopreservation of pre-embryos if the genetic test takes several days (such methods are now available and would enable unaffected pre-embryos to be transferred to the patient at an appropriate time in a subsequent menstrual cycle), and 3) the development of genetic tests that are accurate and sufficiently sensitive for a reliable diagnosis.

Experiments with pre-embryos of laboratory animals suggest that cells could be biopsied either by removing a portion of trophectoderm that is induced to herniate through the overlying zona pellucida or by removing the zona pellucida and dislodging a cell from a cleavage stage pre-embryo at about the eight-cell stage. Recently, Handyside et al. (1989) reported removing a single cell from each of 30 six to ten-cell stage pre-embryos by making a hole in the zona pellucida and removing a cell with a micromanipulator. These three approaches are illustrated in Figure 7 and could be used for obtaining material for preimplantation diagnosis. It may be possible to culture the biopsied cells in order to increase the material available for the test. Several types of genetic test are now being evaluated for use in preimplantation diagnosis. These include chromosome analysis (e.g. Angell et al., 1988), biochemical assays (e.g. Monk, 1988) and molecular tests involving either *in situ* hybridisation (West et al., 1987, 1988; Jones et al., 1987; Penketh et al., 1989) or amplification of target DNA by the polymerase chain reaction (Handyside et al., 1989).

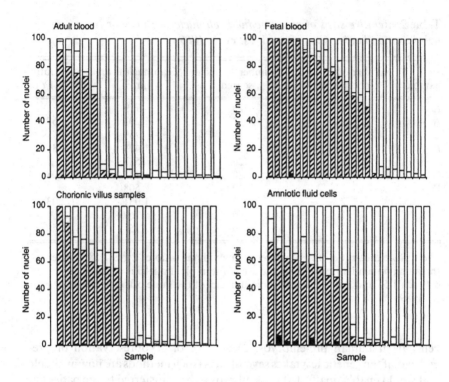

Figure 4. Quantitative *in situ* hybridisation analysis of 100 interphase nuclei from (a) 19 adult lymphocytes, (b) 23 fetal lymphocytes, (c) 20 chorionic villus cell samples, and (d) 20 amniotic fluid cell samples. Shading indicates incidence of nuclei with 0 (☐), 1 equivocal (☐), 1 (▨), or >1 (■) Y hybridisation bodies. Reproduced with permission from West *et al.*, 1989*a*.

Our own work has established that it would be feasible to diagnose the sex of a pre-embryo by DNA–DNA *in situ* hybridisation (West *et al.*, 1987, 1988). Fourteen morphologically normal pre-embryos and nine that appeared abnormal were analysed by *in situ* hybridisation to tritiated Y probe (Figs. 8–10). All 14 morphologically normal pre-embryos (ranging from the two-cell stage to blastocysts with 107 to 156 cells) were sexed with confidence (six females and eight males). Some of the morphologically abnormal pre-embryos had small (probably fragmenting) nuclei without hybridisation bodies and two of these pre-embryos were not sexed. All four male morphologically abnormal, pre-embryos (4–8 cell stages) had one or more nucleus with multiple hybridisation bodies and about 10% of nuclei in the three male blastocysts had two hybridisation bodies. It seems likely that polyploidy occurs both during the normal development of human blatocysts and in early

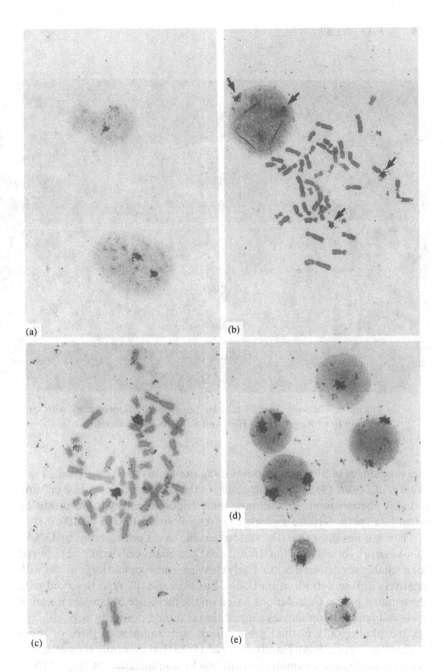

Figure 5. Examples of nuclei with more than one Y hybridisation body: (a) amniotic fluid cell nuclei; (b) cultured lymphocytes from an adult male with a Y/14 translocation: 46,XY,t(Y;14); (c) and (d) cultured lymphocytes from an adult male 47,XYY; (e) blood smear from an adult male 47,XYY. Figure 5(a) is reproduced, with permission, from West et al., 1989a and others are unpublished observations of P. M. Ellis, J. D. West, K. M. West, R. S. Murray and M. C. Coyle (b) and C. M. Gosden, J. D. West, K. M. West, Z. Davidson & C. Davidson (c)–(e).

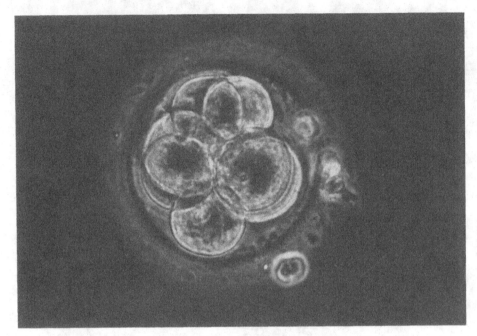

Figure 6. Cleaving (7-cell) human pre-embryo. Reproduced with permission from Angell *et al.*, 1986 and S. Karger AG, Basel.

cleavage stages of abnormally developing pre-embryos (Angell *et al.*, 1987; West *et al.*, 1987, 1988). The polyploidy in the latter group may be a culture-induced phenomenon. In addition, some tripoloid embryos are produced when two sperm fertilise a single egg (Angell *et al.*, 1987; Fig. 10(c)).

These studies show that it would be feasible to sex pre-embryos by DNA–DNA *in situ* hybridisation but it is not clear how many cells would be required for a reliable sex determination. Early cleavage embryos had large nuclei with relatively diffuse hybridisation bodies but were usually readily scored with the tritiated probe. Calculations based on the incidence of labelled nuclei in male and female lymphocytes suggest that it should normally be possible to sex morphologically normal pre-embryos with samples of three or more nuclei (West *et al.*, 1988). Handyside *et al.* (1989) have recently showed that it is also feasible to sex human pre-embryos by amplification of target DNA, from a single blastomere, with the polymerase chain reaction (PCR). Although, in this case, approximately 1500 copies of the target DNA were present in each nucleus, PCR has been used to detect a single copy sequence from one cell (Li *et al.*, 1988). This offers the prospect of an enormous

Sexing the human conceptus by *in situ* hybridisation

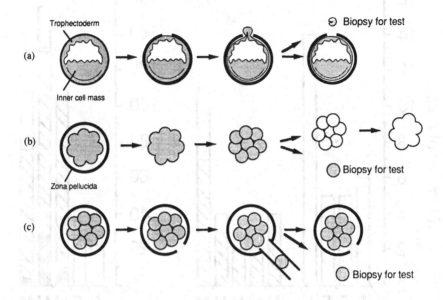

Figure 7. Diagram showing three possible approaches to pre-embryo biopsy. (a) A portion of trophectoderm may be removed after causing it to herniate through a hole made in the zona pellucida. This approach is based on the micromanipulation methods described by Gardner and Edwards (1968) and Papaioannou (1981) and has been used to biopsy marmoset and mouse pre-embryos (Summers *et al.*, 1988; Monk *et al.*, 1988). (b) A single blastomere from a cleavage stage (about the 8-cell stage) pre-embryo could be removed after dissolving the surrounding zona pellucida and, if necessary, decompacting the blastomeres in a medium lacking calcium and magnesium (described for mouse pre-embryos by Monk and Handyside, 1988). (c) A single blastomere from a cleavage stage pre-embryo could be removed with a micropipette, through a hole made in the surrounding zona pellucida, as described for human pre-embryos by Handyside *et al.*, 1989.

improvement in the sensitivity of many of the molecular genetic tests used for prenatal diagnosis.

The development of sensitive genetic tests (notably by PCR or *in situ* hybridisation) makes it likely that pre-implantation diagnosis will be feasible in the near future. This is likely to benefit couples who are at high risk for single-gene defects and anxious to avoid the need for a therapeutic abortion. Pre-implantation diagnosis may be used less frequently for couples at lower risk for genetic disorders, such as many of the spontaneous chromosomal aberrations, where the risk of therapeutic termination is correspondingly lower. (For couples at risk, the risk of conceiving an affected conceptus is

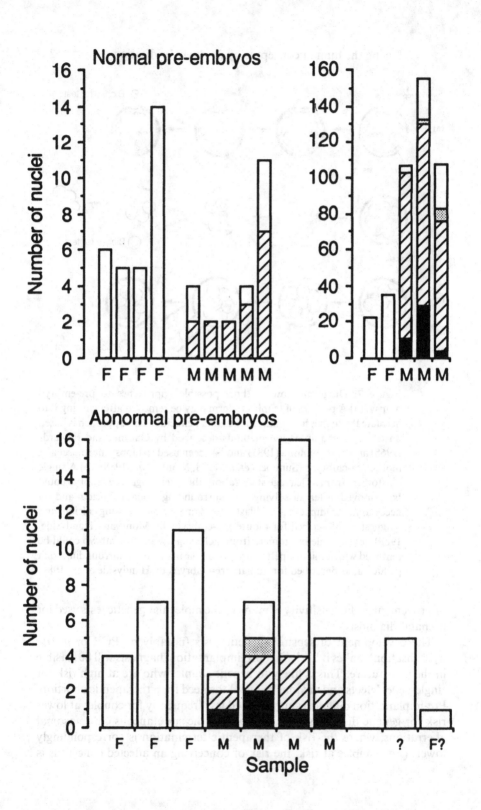

typically 25% for an autosomal recessive, 50% for an autosomal or X-linked dominant and 25% (50% of the males) for an X-linked recessive trait. In contrast, the frequency of the most common chromosomal disorder (trisomy 21) is only 2.42% at term for mothers over the age of 40 and lower still for younger mothers (e.g. Brock, 1982).)

Identification of Y-bearing spermatozoa

In situ hybridisation can also be used to distinguish between X-bearing and Y-bearing spermatozoa (West, West & Aitken, 1989*b*) as shown in Figs 11 and 12. Prior to hybridisation with the Y probe, the sperm heads were decondensed by treating them with ethylenediaminetetra-acetic acid (EDTA), dithiothreitol (DTT) and sodium dodecyl sulphate (SDS). This provides a means of evaluating the various methods that purport to separate X- and Y-bearing spermatozoa. If a woman who is heterozygous for an X-linked genetic disease could be inseminated with only X-bearing spermatozoa, she would avoid the risk of conceiving an affected son without the need for pre-natal or pre-implantation diagnosis.

Polyploid nuclei

The occurrence of nuclei with multiple hybridisation bodies has been discussed in previous sections. It seems likely that those found in the human blastocyst and among amniotic fluid cells (Fig. 5) are polyploid (Angell *et al.*, 1987; West *et al.*, 1987, 1988, 1989*a*). Blastocysts of other mammalian species also contain polyploid cells and in the pig these are primarily localised in the trophectoderm layer (Long & Williams, 1982). It is

Figure 8. Incidence of nuclei with 0 (□), 1 equivocal (▨), 1 (▧), or > 1 (■) Y hybridisation bodies among 14 morphologically normal and 9 morphologically abnormal human pre-embryos. Based on the data of West *et al.*, 1988 with additional data for 3 further pre-embryos (R. R. Angell, J. D. West, K. M. West, L. Harkness, A. F. Glasier, M. W. Rodger & D. T. Baird, unpublished data). The assigned sex is shown as F (female) or M (male). One morphologically normal embryo with two hybridisation bodies in the nuclei originally had 3 pronuclei and may have been an XXY triploid mosaic (West *et al.*, 1987). Two of the abnormal pre-embryos were not sexed. In one case only one cell was recovered and, in the other, all five nuclei were negative but of abnormal morphology (probably fragmented). Small fragmented nuclei without hybridisation bodies were also seen in some of the morphologically abnormal male pre-embryos. Pretreated pre-embryos (e.g. embryo in Fig. 10(c) was pretreated with spermidine *bis*-acridine) are not included in this figure. See West *et al.*, 1987 for details.

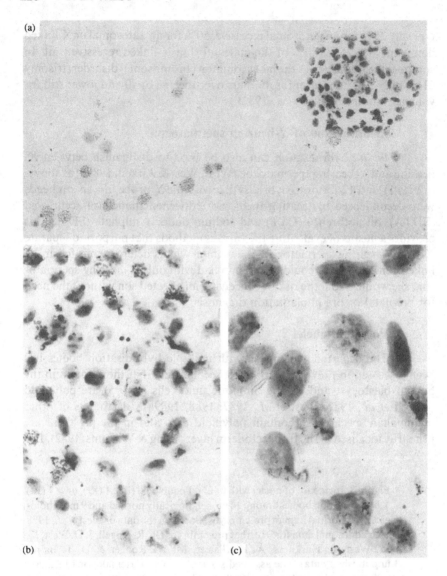

Figure 9. Hybridisation of tritiated Y probe to nuclei of a human blastocyst with 107 nuclei. Low power magnification (a) shows the nuclei spread on the microscope slide and higher magnifications (b), (c) show most nuclei have a single Y hybridisation body and some have two hybridisation bodies.

Sexing the human conceptus by *in situ* hybridisation 221

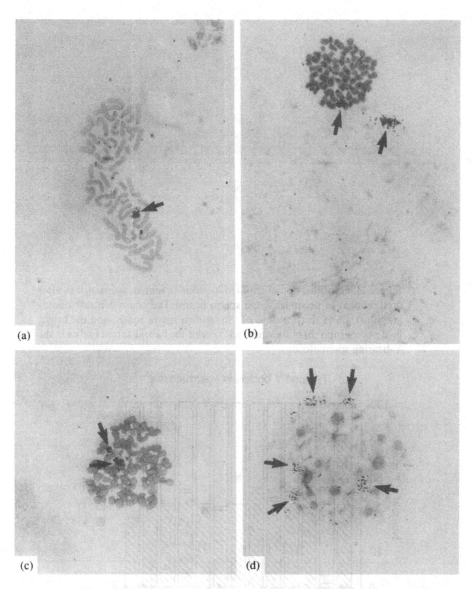

Figure 10. Hybridisation of tritiated Y probe to nuclei of human pre-embryos (Y hybridisation bodies are arrowed): (a) metaphase from a two-cell pre-embryo; (b) metaphase and part of a large interphase nucleus from a six-cell pre-embryo; (c) XYY triploid metaphase from a two-cell embryo that developed from a zygote with three pronuclei; (d) interphase nucleus from morphologically abnormal eight-cell pre-embryo with five Y hybridisation bodies.

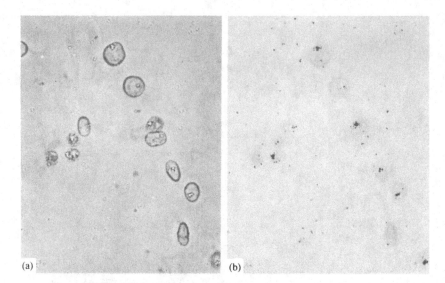

Figure 11. Hybridisation of tritiated Y probe to human spermatozoa after pretreatment to decondense the sperm heads. The phase contrast photograph (a) shows the positions of the swollen sperm heads, and the bright field photograph (b) of the same field reveals the hybridisation bodies in the Y-bearing spermatozoa.

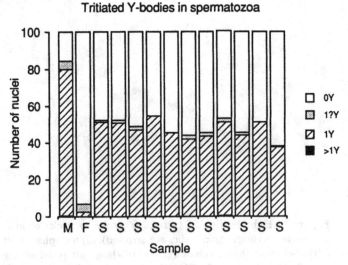

Figure 12. Incidence of nuclei with 0 (□), 1 equivocal (□), 1 (▨), or >1 (■) Y hybridisation bodies among male (M) and female (F) control somatic nuclei and 11 preparations of human spermatozoa (S). Based on data from West et al., 1989b.

tempting to speculate that polyploid nuclei are present in the trophectoderm of the human blastocyst and later derivatives of the trophectoderm, such as the trophoblast cells that are found in amniotic fluid samples. Elucidation of the origin and location of polyploid cells in the developing human conceptus will require the analysis of histological sections. In a preliminary attempt to bridge the gap between the observations on preimplantation stage pre-embryos and cells from the second trimester conceptus, we have used *in situ* hybridisation to look for polyploidy among cells cultured from first trimester fetuses and extraembryonic membranes by *in situ* hybridisation to the Y probe (R. R. Angell, L. Harkness, J. D. West, K. M. West, M. W. Rodger and D. T. Baird unpublished data). However, no samples have yet shown significant frequencies of nuclei with multiple hybridisation bodies. *In situ* hybridisation with a chromosome-specific probe has also been used to identify polyploid interphase nuclei in the mouse liver (Varmuza *et al.*, 1988) and this approach promises to be a valuable experimental tool.

Future use of *in situ* hybridisation for prenatal diagnosis

It is desirable to perform a cytogenetic analysis for prenatal diagnosis of a chorionic villus sample on a direct preparation rather than on cultured material both to avoid overgrowth by any maternal cells and to obtain a rapid diagnosis. However, it is often difficult to obtain sufficient good quality metaphase preparations from direct preparations. *In situ* hybridisation could be done on part a. of sample to provide a rapid sex determination on interphase nuclei while the rest of the sample was used for cytogenetic or other tests. Similarly, tests involving *in situ* hybridisation can be done on nucleated cells present in blood smears without the need to culture lymphocytes (Fig. 5(e); J. D. West and K. M. West unpublished data). In the future it may also be possible to use *in situ* hybridisation on trophoblast cells isolated from the maternal circulation (Covone *et al.*, 1984; Kozma *et al.*, 1986). This would provide prenatal diagnosis without the need for invasive biopsy of the conceptus but there is a danger that some trophoblast cells may have persisted from a previous (possibly undetected) pregnancy and produce an inaccurate diagnosis.

The need for only interphase nuclei means that prenatal diagnosis by *in situ* hybridisation can begin without a preliminary period of culture but it would also be helpful if the time taken to perform the test could be reduced to a minimum. The work with the Y probe, described above, produced reliable results but the autoradiography took six days. This time can be considerably reduced by using non-radioactively labelled DNA probes. The guanosine nucleotides of the DNA probe can be modified by treatment with 2-acetylaminofluorene (AAF) as described by Landegent *et al.* (1984) and then the modified probe can be detected by immunocytochemical methods. More

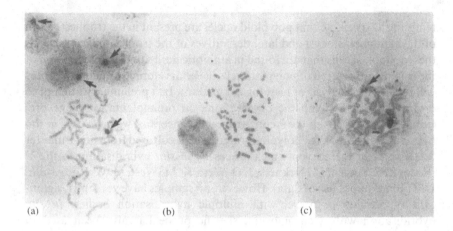

(a) (b) (c)

Figure 13. Hybridisation of biotinylated Y probe to (a) male control lymphocytes, (b) female control lymphocytes and (c) mitotic chromosomes from an eight-cell human pre-embryo. The hybridisation bodies (arrowed) are seen as brown precipitates after histochemical staining for horseradish peroxidase. (The large dark mass in (c) is stain debris.)

commonly the probe can be labelled by incorporating biotinylated nucleotides and recently nucleotides labelled with digoxigenin (digoxigenin-dUTP) have become commercially available (from Boehringer Mannheim) as an alternative to biotin. The labelled probes are detected using either enzyme-conjugated or fluorochrome-conjugated antibodies. Biotinylated probes may be detected with avidin or streptavidin instead of anti-biotin antibodies. These non-radioactive *in situ* hybridisation methods are extensively discussed elsewhere in this volume.

Burns *et al.* (1985) were among the first to use biotinylated Y probe to sex interphase cells by *in situ* hybridisation and we have used a similar horseradish peroxidase immunocytochemical system to detect the biotinylated probe. Figure 13(a), (b) illustrates male and female control lymphocytes after hybridisation to biotinylated Y probe and immunohistochemical staining for horseradish peroxidase. The frequency of hybridisation bodies among such controls is shown in Figure 14. We have also applied this technique to sex a small number of human pre-embryos (Fig. 13(c) and West *et al.*, 1988) in order to test the feasibility of this as a test for pre-implantation diagnosis. Although we found that this method was capable of discriminating between male and female control samples when 100 interphase nuclei were scored, we obtained more variable results with pre-embryos. Early pre-embryos have only a few cells and, after hybridisation

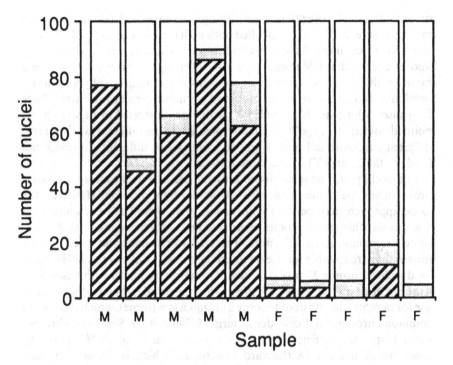

Figure 14. Incidence of nuclei with 0 (□), 1 equivocal (▨), 1 (▨), or >1 (■) Y hybridisation bodies from control slides of male (M) and female (F) lymphocytes hybridised to biotinylated Y probe. Reproduced with permission from West et al., 1988.

with biotinylated Y probe, the hybridisation bodies in the large nuclei of early cleavage stages were often diffuse and less obvious than with tritiated Y probe. Penketh et al. (1989) have used alkaline phosphatase immunocytochemistry as an end-point for identifying biotinylated Y probe hybridised to human pre-embryos and Pinkel et al. (1986a, b) have reported a very sensitive fluorescent immunochemical means of detecting biotinylated DNA probes. This fluorescent method has been used by several laboratories for fetal sexing (e.g. Kozma & Adinolfi, 1988; Guyot et al., 1988).

Although non-radioactive labelling with biotin allows diagnosis of sex in 1 or 2 days instead of the 8 days normally used with tritiated probe, the technique suffers from some disadvantages. In our hands, at least, the biotin/horseradish peroxidase system proved less sensitive than tritium and produced more variable results and, in some cases, a higher level of background staining. The fluorescent end-point is not permanent which may be a drawback for detailed analysis or when slides are stored for later reference. In order to try to overcome these disadvantages we have used a double labelling method where the Y probe is labelled both with tritium and biotin (J. D. West & M. Keighren, unpublished). Hybridisation bodies are initially detected by fluorescence with a UV microscope after overnight hybridisation to give a rapid result. The slides are then processed for autoradiography and developed after 6 days' exposure to provide a permanent record. Control slides hybridised with a double labelled Y probe are illustrated in Figure 15. As pointed out earlier (page 208), it is also possible to use different labels for two different chromosomal probes and so identify two different chromosomes (such as the X and Y) in the same nucleus.

In this chapter, I have discussed only the use of a Y chromosomal DNA probe to sex the human conceptus by *in situ* hybridisation. However, the same approach may be used with probes that hybridise to reiterated sequences on other chromosomes in order to detect specific numerical chromosome anomalies such as trisomies. Table 1 lists chromosome specific DNA probes that are suitable for use on interphase nuclei. Julien *et al.* (1986) have used a chromosome 21 probe to identify interphase nuclei with trisomy 21 (three hybridisation bodies as distinct from the usual two) and discussed its use for prenatal diagnosis of trisomy 21 (Down syndrome), which is the most common chromosomal disorder at term (1/700 births). Similarly, chromosome 18 probes (Cremer *et al.*, 1986; Vorsanova *et al.*, 1986) allow the detection of trisomy 18 (Edward syndrome), which is the second most common trisomy at term (1/3000 births). In some interphase nuclei, the distinction between 2 and 3 hybridisation bodies may be less obvious than the distinction between 0 and 1 that is used for sexing with the Y probe (see pages 208–11). However, if sufficient interphase nuclei are scored, these chromosome aberrations should be readily detected and *in situ* hybridisation is likely to prove to be a valuable tool for prenatal diagnosis.

Protocols

Most of the work described in this chapter involves *in situ* hybridisation to tritiated Y probe and this method is detailed below. We have also used three methods to detect the hybridised biotinylated Y probe. Initially we used goat anti-biotin antiserum followed by horseradish peroxidase-conjugated rabbit anti-goat and histochemical staining for horseradish peroxidase activity. This method is described elsewhere (Manuelides, 1985; Burns *et al.*, 1985; West *et al.*, 1988). More recently we have used the ABC kit from Vector labs (horseradish peroxidase-conjugated avidin, followed by biotinylated anti-avidin and finally avidin/biotinylated horseradish peroxidase complex and histochemical staining for horseradish peroxidase activity) and the similar fluorescent method described by Pinkel *et al.*, 1986*a*, *b*. We have used the fluorescent method alone and in combination

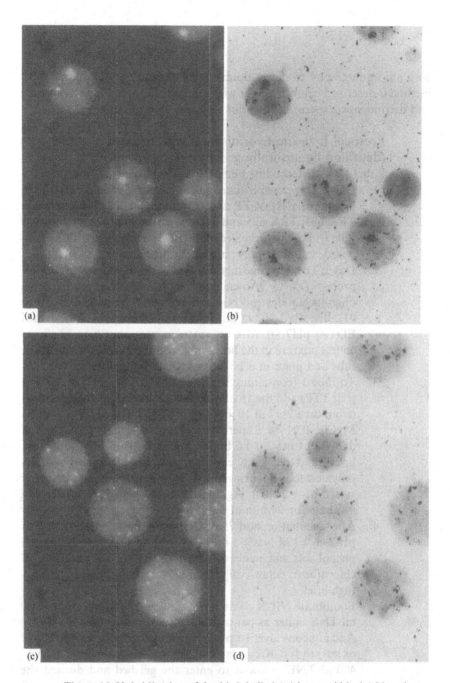

Figure 15. Hybridisation of double labelled (tritium and biotin) Y probe to control male (a) and (b) and female (c) and (d) lymphoblastoid cell lines. First (a) and (c) U.V. microscopy was used to detect the biotin (after several sequential treatments with fluorescein-conjugated avidin and biotinylated anti-avidin), then (b) and (d) autoradiography was used to detect the tritium in the same probe hybridised to the same nuclei. Both detection methods revealed the Y hybridisation bodies in the male cells (a) and (b).

with autoradiography to detect double labelled probes. The double label method is described below together with methods for preparing cell spreads and decondensing sperm heads prior to *in situ* hybridisation.

Protocol I: *In situ* hybridisation with tritiated Y probe (detection by autoradiography)
Label DNA (Method of Feinberg & Vogelstein 1983)
1. Day 1: Dry down tritiated nucleotides (0.5 nmole [^{3}H] dATP, 0.5 nmole [^{3}H] dCTP and 0.5 nmole [^{3}H] TTP (Amersham TRK 633, TRK 625 and TRK 576)) overnight in a microfuge tube (tube 1) with a stream of air in a designated fume cupboard, taking appropriate safety precautions.
2. Day 2: To microfuge tube 2 add 12 μl sterile water, 2 μl (1 μg) Y probe DNA (Amersham RPN 1305X) and 2 μl oligonucleotides (pd(N)6 sodium salt (Pharmacia 27–2166–01), dissolved (1 mg/ml) in TE buffer (10 mM Tris, 1 mM EDTA; pH7.5)). Mix on vortex mixer, centrifuge briefly to collect mixture in the bottom of the tube, pierce the top of the tube and place in a boiling water bath for 3 min.
3. To tube 1 (containing the dried [^{3}H] dATP, [^{3}H] dCTP and [^{3}H] TTP) add the 16 μl of boiled DNA plus oligonucleotides (from tube 2), 2 μl 10 × NT salts (0.5 M Tris pH 7–8, 0.05 M MgCl$_2$, 0.01 M β mercaptoethanol, 0.5 mg/ml BSA), 1 μl dGTP (0.1 mM in TE buffer; Sigma D4010). Mix, centrifuge briefly and add 1μl Klenow enzyme (large fragment of DNA polymerase I; Boehringer 104523; 5 units/μl in 50% glycerol: stored at -20°C.). Agitate the tube, centrifuge briefly and incubate for 90 min at 37°C.
4. After incubation add 75μl TNE (column buffer: 0.01 M Tris pH 7–8, 0.2 M NaCl, 1 mM EDTA) and 50μl water-saturated phenol, mix and incubate for 10 min at 37°C. Then add 50μl of chloroform: octan-2-ol mixture (24:1 ratio), mix and centrifuge briefly.
5. Equilibrate NICK column (Pharmacia 17-0855-01) with 9–10 ml TNE buffer as per manufacturer's instructions (3 × 3 ml). Add aqueous layer (top layer) from phenol extraction (100 μl or less) to NICK column and allow to enter the gel bed. Add 400 μl TNE, allow it to enter the gel bed and discard the displaced buffer. Place a new, sterile microfuge tube (tube 3) under the column and add another 400 μl TNE. Collect purified labelled DNA in tube 3 (in eluate of second 400 μl of TNE). If required count 2 μl from discard tube and 2 μl from DNA

fraction, in a scintillation counter, to confirm the presence of labelled DNA in tube 3.
6. To the DNA fraction in tube 3, add 10 µl (100 µg) of carrier DNA (salmon sperm DNA, Sigma D1626: 10 mg/ml in TE buffer, sonicated and stored at 4 °C) and 40 µl 3 M ammonium acetate. Mix, add 1 ml absolute ethanol, mix gently and place at −20 °C for 1 h or more.
7. Centrifuge at high speed in a microcentrifuge (e.g. 13 000 rpm in an MSE microcentaur centrifuge is 11 600 × g) for 10 min. Pour off supernatant (count 50 µl in a scintillation counter, if required, to check that the supernatant has only a low level of tritium) and dry the pellet in a desiccator. Dissolve the pellet in approximately 100 µl of sterile, distilled water. The minimum volume of water is calculated as ((number of slides +1) × 4) + 1 µl (e.g. for 18 slides dissolve the probe in 77 µl water). Excess labelled probe can be stored frozen for several weeks, if required. Count the tritium in 1 µl of the dissolved probe with a scintillation counter. This is approximately one-quarter of the counts per slide – see also step 12 below.

Hybridisation (Method of Gosden, Middleton, Rout & De Angelis, 1986)
8. Dilute 1 ml of ribonuclease stock in 250 ml 2 × SSC in staining trough (40 µg/ml in 2 × SSC) and equilibrate in a 37 °C water bath. (Ribonuclease stock is 10mg/ml ribonuclease A (Sigma R5125) in 2 × SSX, boiled for 5–10 min and stored at −20 °C; SSC solutions are diluted from a 20 × SSC stock (3 M NaCl, 0.3 M sodium citrate, pH 7–7.4)). Select microscope slides with appropriate cell spreads and incubate them in ribonuclease for 1 h at 37 °C and dehydrate through graded alcohols (2 min in each of 70%, 90% and 100% ethanol) and air dry.
9. Equilibrate 250 ml of 70% formamide, 0.6 × SSC to 70 °C in a staining trough in a water bath and place the slides in this for 2 min to denature the target DNA. Dehydrate through graded alcohols (as above) and air dry.
10. Prepare and denature the DNA probe. (The final volume should be sufficient for 20 µl per slide and is five times the volume of the DNA remaining from step 7, above. For 18 slides we use 380 µl, allowing 360 µl for the slides and 20 µl spare for tritium counts). To tube 3 containing DNA probe in water (from step 7) representing 20% of the final volume add 10 × SSCP (20% of total), tRNA carrier (10% of total) and

formamide with 20% dextran sulphate (50% of total). Mix thoroughly, incubate the tube in a 70 °C water bath for 5 min to denature the probe and cool on ice. (10 × SSCP:1.2 M NaCl, 0.15 M sodium citrate, 0.2 M $NaPO_4$; tRNA (Sigma R1753) 4 mg/ml in TE buffer stored at -4°C.)

11. For each slide, add 20 μl of the denatured probe mix to a cleaned coverslip and apply the slide to the coverslip. Seal the edge of the coverslip with a rubber adhesive (e.g. Tip Top, Holdtite or Weldtite), allow this to dry and incubate at 37 °C overnight.
12. Count tritum levels in 20 μl of the probe mix (if spare) with a scintillation counter to estimate the counts per slide.
13. Prepare and equilibrate wash buffers to 39 °C for use on day 2 (1 litre of 1 × SSC; 1 litre of 50% formamide, 1 × SSC).
14. Day 2: Wash off non-specific hybridisation. Remove coverslips and rubber solution to radioactive waste. Wash slides in 50% formamide 1 × SSC (4 × 5 min) then wash in 1 × SSC (4 × 5 min) at 39 °C. Dehydrate through graded alcohols (2 min in each of 70%, 90% and 100% ethanol) and air dry.

Note: Formamide in the hybridisation mix provides stringent hybridisation conditions and prevents random hybridisation of incorrect sequences. Dextran sulphate excludes water from the DNA and so increases the effective solute concentration and speeds up the rate of hybridisation. Formamide in the washes removes any DNA probe that has stuck randomly to the slide.

Autoradiography (Method of Gosden, Middleton, Rout & De Angelis 1986)

15. In a dark room illuminated by a safelight, prepare 14 ml of emulsion by dissolving gel-form photographic emulsion (e.g. Ilford L4) in 7 ml of sterile water (in a 25 ml measuring cylinder warmed to 45 °C in a water bath) until the total volume is 14 ml. (Handle the solid emulsion with autoclaved plastic forceps.) Mix the emulsion gently and pour it into a dipping dish and stand the dipping dish in water in a beaker placed in the 45 °C water bath. (Plastic slide holders designed to carry two microscope slides can be used as dipping dishes or glass dipping dishes can be made to order by R. B. Radley & Co. Ltd., Shire Hill, Saffron Waldon, Essex, CB11 3AZ.)
16. Dip slides in the warm emulsion (two slides at a time, back to back) and stand the slides in a rack to drain. Place the rack in a light-tight cupboard or box for 3 h to dry then transfer the slides to a slide rack in light-box, containing silica gel. Store the

box at 4°C away from any source of radiation to expose the autoradiograph. (We routinely expose slides hybridised with probe pHY2.1 for 6 days but normally test slides should be developed at intervals to determine the optimum exposure.)
17. Develop autoradiographs (after exposing for an appropriate period). In a dark room illuminated by a safelight, place the slides (in a staining rack) in a dish of photographic developer (Kodak D19) for 5 min, and agitate the rack gently every minute. Rinse briefly in a dish of distilled water and then transfer to a dish of fixative (Kodak Unifix diluted 1:3 with distilled water) for 5 min, and agitate the rack gently every minute. (The lights may then be switched on.) Place the rack in water then transfer it to running tap water for 6–10 min. Place the slides in a clean rack to air dry.
18. Counterstain. Put up to five slides in an empty Coplin jar and pour in freshly prepared Giemsa stain (2.5 ml Gurr's improved R66 Giemsa solution plus 47.5 ml Gurr's buffer, pH 7.0). After 5 min, place the Coplin jar in a sink and pour in Gurr's buffer until the top layer of stain has floated off (to avoid deposits of stain debris on slide). Pour off the remaining stain solution and rinse the slides in tap water. Dry the slides in a rack and mount them with DPX under a clean coverslip.

Protocol II: *In situ* hybridisation with double-labelled Y probe (detection by fluorescence microscopy followed by autoradiography)
Label DNA
1. Day 1: Dry down tritiated nucleotides (0.5 nmole [^{3}H] dATP and 0.5 nmole [^{3}H] dCTP) overnight in a microfuge tube (tube 1).
2. Day 2: To tube 2 add: 9.5 µl sterile water, 2.0 µl (1 µg) DNA probe and 2 µl oligonucleotides. Mix on vortex mixer, centrifuge briefly to collect mixture in the bottom of the tube, pierce the top of the tube and place in a boiling water bath for 3 min.
3. To tube 1 (with dried [^{3}H] nucleotides) add: 13.5 µl boiled DNA and oligonucleotides, 2.0 µl 10 × NT salts, 2.5 µl biotinylated dUTP and 1.0 µl dGTP (0.1 mM). Mix, centrifuge briefly and add 1µl Klenow enzyme (Boehringer 104523; 5 units/µl). Agitate the tube, centrifuge briefly and incubate for 90 min at 37°C.
4. After incubation add 5 µl stop buffer (300 mM Na$_2$EDTA pH8.0, filter sterilised), 1 µl 5% SDS (sodium dodecyl sulphate) and 25 µl TNE column buffer.

5. Load onto a pre-equilibrated NICK column and elute as in step 5 of Protocol I. The double labelled DNA should be in the second 400 μl of TNE eluted from the column. Check tritium levels in 2 μl of the discard fraction and 2 μl of the sample fraction with a scintillation counter if required.
6. To the DNA fraction in tube 3, add 10 μl (100 μg) of carrier salmon sperm DNA and 40 μl 3 M ammonium acetate. Mix well, add 1 ml absolute ethanol, mix gently and place at $-20\,°C$ for 1 h or more.
7. Centrifuge at high speed in a microcentrifuge for 10 min, pour off supernatant and dry the pellet in a desiccator. Dissolve the pellet in 100 μl TE buffer and store at $-20\,°C$.

Hybridisation (See Protocol above for further details.)

8. Incubate slides in ribonuclease (40 μg/ml in 2 × SSC) for 1 h at 37°C. Dehydrate through graded alcohols (2 min in each of 70%, 90% and 100% ethanol) and air dry.
9. Denature (70% formamide; 0.6 × SSC) for 2 min at 70°C. Dehydrate through graded alcohols (as above) and air dry.
10. Incubate in proteinase K solution (60 ng/ml in 20 mM Tris pH 7.5; 2 mM $CaCl_2$) for 7 min at 37°C. (Use 15 μl of 1 mg/ml stock solution.)

Note: Proteinase K treatment improves the detection of target DNA in interphase nuclei and some older metaphase preparations.

11. Prepare and denature the DNA probe (see Protocol I, steps 8–14 for further details):

DNA probe (from tube 3 in step 6)	20% of final volume
10 × SSCP	20%
tRNA (carrier)	10%
formamide plus dextran sulphate	50%

    Mix thoroughly, incubate the tube in a 70°C water bath for 5 min to denature the probe and cool on ice.
12. For each slide, add 20 μl of the denatured probe mix to a cleaned coverslip and apply the slide to the coverslip. Seal the edge of the coverslip with a rubber adhesive, allow this to dry and incubate at 37°C overnight.
13. Prepare and equilibrate wash buffers to 39°C for use on day 2 (1 litre of 1 × SSC; 1 litre of 50% formamide, 1 × SSC).
14. Day 2: Remove coverslips and rubber solution and wash slides in 50% formamide 1 × SSC (4 × 5 min) then wash in 1 × SSC (4 × 5 min) at 39°C. Immerse slides in 4 × SSC, 0.05% Triton X-100 (pre-heated in 39°C water bath).

Biotin detection with fluorescent avidin (Method of Pinkel et al., 1986 a, b)

15. Treat slides with blocking buffer (4 × SSC; 0.05% Triton X-100; 5% low fat milk powder). Use 30 μl per slide for 5 min at room temperature under cover slips. (Equilibrate blocking buffer to 37°C before use.)
16. Remove coverslips and retain them for later use. Drain excess fluid and apply first antibody: FITC-avidin DCS (Vector A2011), diluted 1 in 500 in PBS. (Keep this antiserum in the dark when not in use and centrifuge it before dilution to remove any immune complex from the supernatant.) Use 30 μl/slide and incubate at 37°C for 30 min in a humid container.
17. Wash 3 × 2 min in 4 × SSC, 0.05% Triton X-100 at 42°C (agitate slides every minute).
18. Drain excess fluid and apply second antibody: biotinylated goat anti-avidin (Vector BA0300) diluted 1:100 in PBS. Use 30 μl/slide and clean coverslips. Incubate at 37°C for 30 min in a humid container.
19. Wash 3 × 2 min in 4 × SSC, 0.05% Triton X-100 at 42°C (agitate slides every minute).
20. Repeat step 16 using fresh coverslips.
21. Wash 3 × 2 min in 4 × SSC, 0.05% Triton X-100 at 42°C (agitate slides every minute).
22. Repeat steps 18 to 21 inclusive if required.
23. Mount slides in Citifluor AF1 glycerol-phosphate buffered solution (Citifluor Ltd, London) containing 1.5 μg/ml propidium iodide. Use 40 μl/slide and seal with rubber solution. Keep slides in the dark until viewed by UV microscopy. The chromosomes and nuclei fluoresce orange and the fluorescein-labelled Y-probe appears bright yellow.
24. After viewing the slides under fluorescent microscope wash in 1 × PBS at 37°C (1 × 3 min), then in 2 × SSC at 37°C (4 × 3 min). Dehydrate through graded alcohols (2 min in each of 70%, 90% and 100% ethanol) and air dry.
25. Detect tritiated Y probe by autoradiography, as Protocol I, steps 15–18 above.

Protocol III: Preparation of cell spreads from cell lines for in situ hybridisation

1. Add 0.3 ml colcemid (10 μg/ml) to a flask containing 15 ml of cells in culture medium and leave in 37°C incubator for 1–3 hours. Warm 0.08 M KCl to 37°C.

2. Centrifuge cells (1000 rpm (175 × g) for 5 min) and pour off the supernatant. Resuspend cell pellet in 0.08 M KCl and leave at 37°C for 10–15 min. Centrifuge again, pour off supernatant and vortex gently to mix cells.
3. Add two Pasteur pipette fulls of freshly prepared 3:1 fixative methanol:acetic acid), dropwise, while holding the tube on a vortex mixer. Centrifuge and pour off the supernatant and vortex gently.
4. Repeat step 3 two more times.
5. Add 1 ml of fixative, mix well and drop one drop onto each clean, labelled slide. Air dry, check under phase contrast microscopy and store slides in racks in a clean dry box.

Protocol IV: Decondensation of sperm heads prior to *in situ* hybridisation

1. To 1 ml of washed sperm in BWW medium (Biggers, Whitten & Whittingham, 1971), add 1 ml of 6 mM EDTA (60 μl 0.2 M EDTA plus 1940 μl BWW). Leave at room temperature for 5–10 min and centrifuge (1000 rpm (175 × g) for 5 min).
2. Discard supernatant and resuspend sperm in 1 ml of 2 mM DTT (154 μl of 4 g/ml stock dithiothreitol plus 1846 μl BWW medium). Leave at room temperature for 45 min and centrifuge as before.
3. Discard supernatant, resuspend sperm in BWW medium and centrifuge.
4. Add one Pasteur pipette full of freshly prepared 3:1 fixative (methanol:acetic acid), dropwise, while holding the tube on a vortex mixer. Leave at room temperature for 30–60 min and centrifuge.
5. Discard the supernatant and add 3:1 fixative (eg1ml) dropwise, while holding the tube on a vortex mixer. Drop one drop onto each clean, labelled slide. Air dry, check under phase contrast microscopy and store slides in racks in a clean, dry box.
6. Immediately before *in situ* hybridisation place slides flat on a rack and add 0.05% SDS (20 μl of 5% SDS per 2 ml BWW medium). Leave for 5 min, tip off SDS and wash with 2 × SSC. Dehydrate through graded alcohols (2 min in each of 70%, 90% and 100% ethanol) and air dry.
7. After *in situ* hybridisation, stain slides in 0.5% Eosin yellow (BDH, C145380) for 5 min. Rinse in tap water and air-dry. Mount in DPX.

Acknowledgements

I am grateful to many scientific, clinical and technical collaborators for their input into the Y probe work. These include J. R. Gosden, R. R. Angell, C. M. Gosden, K. M. West, M. Keigren, L. Harkness, S. S. Thatcher, A. F.Glasier, M. W. Rodger, D. T. Baird, N. Hastie, H. J. Evans, P. M. Ellis, R. S. Murray, M. C. Coyle and E. Grace. I also thank T. McFetters and E. Pinner for preparing the illustrations, K. M. West and M. Keighren for reading the manuscript, R. R. Angell and S. Karger AG, Basel for permission to reproduce Figure 6 and the Wellcome Trust for financial support.

References

Angell, R. R., Hillier, S. G., West, J. D., Glasier, A. F., Rodger, M. W. & Baird, D. T. (1988). Chromosome anomalies in early human embryos. *Journal of Reproduction and Fertility Supplement*, **36**, 73–81.

Angell, R. R., Sumner, A. T., West, J. D., Thatcher, S. S., Glasier, A. F. & Baird, D. T. (1987). Post fertilization polyploidy in human preimplantation embryos fertilized *in vitro*. *Human Reproduction*, **2**, 721–7.

Biggers, J. D., Whitten, W. K. & Whittingham, D. G. (1971). The culture of mouse embryos *in vitro*. In *Methods in Mammalian Embryology*, ed. Daniel, J. C. Jr. San Francisco: Freeman pp. 86–116.

Brock, D. J. H. (1982). *Early Diagnosis of Fetal Defects*. Churchill Livingstone, Edinburgh, London, Melbourne & New York. 165pp.

Burns, J., Chan, V. T. W., Jonasson, J. A., Fleming, K. A., Taylor, S. & McGee, J. O. D. (1985). Sensitive system for visualising biotinylated DNA probes hybridised in situ: rapid sex determination of intact cells. *Journal of Clinical Pathology*, **38**, 1085–92.

Buster, J. E., Bustillo, M., Rodi, I. A., Cohen, S. W., Hamilton, M., Simon, J. A., Thornycroft, I. H. & Marshall, J. R. (1985). Biologic and morphologic development of donated human ova recovered by nonsurgical uterine lavage. *American Journal of Obstetrics and Gynecology*, **153**, 211–17.

Cook, H. J., Schmidtke, J. & Gosden, J. R. (1982). Characterisation of a human Y chromosome repeated sequence and related sequences in higher primates. *Chromosoma*, **87**, 491–502.

Covone, A. E., Mutton, D., Johnson, P. M. & Adinolphi, M. (1984). Trophoblast cells in peripheral blood from pregnant women. *Lancet*, **ii**, 842–3.

Cremer, T., Landegent, J., Bruckner, A., Scholl, H. P., Schardin, M., Hager, H. D., Devilee, P., Pearson, P. & van der Ploeg, M. (1986). Detection of chromosome aberrations in the human interphase nucleus by visualization of specific target DNAs with radioactive and non-radioactive *in situ* hybridization techniques: diagnosis of trisomy 18 with probe L1.84. *Human Genetics*, **74**, 346–52.

Disteche, C. M., Brown, L., Saal, H., Friedman, C., Thuline, H. C., Hoar, D. I., Pagon, R. A. & Page, D. C. (1986). Molecular detection of a translocation (Y; 15) in a 45X male. *Human Genetics*, **74**, 372–7.

Fantes, J., Gosden, J. R. & Piper, J. (1988). Use of an alphoid satellite sequence to automatically locate the X chromosome, with particular reference to the identification of the fragile X. *Cytogenetic Cell Genetics*, **48**, 142–7.

Feinberg, A. P. & Vogelstein, B. (1983). A technique for radiolabelling DNA restriction endonuclease fragments to high specific activity. *Analytical Biochemistry*, **132**, 6–13.

Gardner, R. L. & Edwards, R. G. (1968). Control of the sex ratio at full term in the rabbit by transferring sexed blastocysts. *Nature (London)*, **218**, 346–8.

Gosden, C. M. (1983). Amniotic fluid cell types and culture. *British Medical Bulletin*, **39**, 348–54.

Gosden, J. R. & Gosden, C. M. (1985). Recombinant DNA technology in prenatal diagnosis. In *Oxford Reviews of Reproductive Biology* (ed. J. R. Clarke), **7**, pp. 73–117.

Gosden, J. R., Gosden, C. M., Christie, S., Cooke, H. J., Morsman, J. M. & Rodeck, C. H. (1984). The use of cloned Y-chromosome-specific DNA probes for fetal sex determination in first trimester prenatal diagnosis. *Human Genetics*, **66**, 347–51.

Gosden, J. R., Middleton, P. G., Rout, D. & De Angelis, C. (1986). Chromosome localisation of the human oncogene ERBA2. *Cytogenetic Cell Genetics*, **43**, 150–3.

Guyot, B., Bazin, A., Sole, Y., Julien, C., Daffos, F. & Forestier, F. (1988). Prenatal diagnosis with biotinylated chromosome specific probes. *Prenatal Diagnosis*, **8**, 485–93.

Handyside, A. H., Pattinson, J. K., Penketh, R. J. A., Dehanty, J. D. A., Winston, R. M. L. & Tuddenham, E. G. D. (1989). Biopsy of human preimplantation embryos and sexing by DNA amplification. *Lancet*, **i**, 347–9.

Jones, K. W., Singh, L. & Edwards, R. G. (1987). The use of probes for the Y chromosome in preimplantation embryo cells. *Human Reproduction*, **2**, 439–45.

Julien, J. C., Bazen, A., Guyot, B., Forestier, F. & Daffos, F. (1986). Rapid prenatal diagnosis of Down's syndrome with *in situ* hybridisation of fluorescent DNA probes. *Lancet*, **ii**, 863–4.

Kozma, R. & Adinolfi, M. (1988). *In situ* fluorescence hybridisation of Y translocations: Cytogenetic analysis using probes Y190 and Y431. *Clinical Genetics*, **33**, 156–61.

Kozma, R., Spring, J., Johnson, P. M. & Adinolphi, M. (1986). Detection of syncytiotrophoblast in maternal peripheral and uterine veins using a monoclonal antibody and flow cytometry. *Human Reproduction*, **1**, 335–6.

Landegent, J. E., Jansen In De Wal, N., Baan, R. A., Hoeijmakers, J. H. J. & Van Der Ploeg, M. (1984). 2-Acetylaminofluorene-modified probes for the indirect hybridocytochemical detection of specific nucleic acid sequences, *Experimental Cell Research*, **153**, 61–72.

Lau, Y. F., Ying, K. L. & Donnell, G. N. (1985). Identification of a case of Y:18 translocation using a Y-specific repetitive DNA probe. *Human Genetics*, **69**, 102–5.

Li, A., Gyllensten, U. B., Cui, X., Saiki, R. K., Erlich, H. A. & Arnheim, N. (1988). Amplification and analysis of DNA sequences in single human sperm and diploid cells, *Nature (London)*, **335**, 414–19.

Long, S. E. & Williams, C. V. (1982). A comparison of the chromosome complement of inner cell mass and trophoblast cells in day 10 pig embryos. *Journal of Reproduction and Fertility*, **66**, 645–8.

Manuelides, L. (1985). *In situ* detection of DNA sequences using biotinylated probes. *Focus*, **7**, 4–8. Bethesda Research Laboratories.

McLaren, A., (1985) Prenatal diagnosis before implantation: opportunities and problems. *Prenatal Diagnosis*, **5**, 85–90.

McLaren, A. (1986) Prelude to embryogenesis. In *Human Embryo Research. Yes or No?* Ciba Foundation, pp. 5–23. London: Tavistock Publications.

Millar, D. S., Davis, L. R., Rodeck, C. H., Nicolaides, K. H. & Mibashan, R. S. (1985). Normal blood cell values in the early mid-trimester fetus. *Prenatal Diagnosis*, **5**, 367–73.

Monk, M. (1988). Preimplantation diagnosis. *BioEssays*, **8**, 184–9.

Monk, M. & Handyside, A. H. (1988). Sexing of preimplantation mouse embryos by measurement of X-linked gene dosage in a single blastomere. *Journal of Reproduction and Fertility*, **82**, 365–8.

Monk, M., Muggleton-Harris, A. L., Rawlings, E. & Whittingham, D. G. (1988). Pre-implantation diagnosis of HPRT-deficient male and carrier female mouse embryos by trophectoderm biopsy. *Human Reproduction*, **3**, 377–81.

Mullis, K. B. & Faloona, F. A. (1987). Specific synthesis of DNA *in vitro* via a polymerase-catalysed chain reaction. *Methods in Enzymology*, **155**, 335–50.

Nicolaides, K. H., Soothill, P. W., Rodeck, C. H. & Campbell, S. (1986). Ultrasound-guided sampling of umbilical cord and placental blood to assess fetal wellbeing. *Lancet*, **i**, 1065–7.

Old, J. M. (1986). Fetal DNA analysis. In (ed. Davis, K. E.), *Human Genetic Diseases. A Practical Approach.* pp. 1–17, Oxford: IRL Press.

Page, D. C., Mosher, R. Simpson, E. M., Fisher, E. M. C., Mardon, G. Pollack, J., McGillivray, B., de la Chapelle, A. & Brown, L. G. (1987). The sex-determining region of the human Y chromosome encodes a finger protein. *Cell*, **51**, 1091–104.

Papaioannou, V. E. (1986). Microsurgery and micromanipulation of early mouse embryos. In *Techniques in Cellular Physiology – Part 1*. pp. 1–27. Amsterdam: Elsevier/North Holland.

Penketh, R. & McLaren, A. (1987). Prospects for prenatal diagnosis during preimplantation human development. In *Fetal Diagnosis of Genetic Defects*. Baillieres International Clinical Obstetrics and Gynaecology vol. 1, pp. 747–64.

Penketh, R. J. A., Delhanty, J. D. A., Vanden Barghe, J. A. et al (1989). Rapid sexing of human embryos by non-radioactive *in situ* hybridisation: potential for preimplantation diagnosis of X-linked disorders. *Prenatal Diagnosis*, **9**, 489–500.

Pinkel, D., Gray, J. W., Trask, B., Van Den Engh, G. Fuscoe, J. & Van Dekken, H. (1986a). Cytogenetic analysis by *in situ* hybridization with fluorescently labeled nucleic acid probes. *Cold Spring Harbor Symposia on Quantitative Biology*, **51**, 151–7.

Pinkel, D., Straume, T. & Gray, J. W. (1986b). Cytogenetic analysis using quantitative high sensitivity, fluorescence hybridization. *Proceedings of the National Academy of Sciences, USA*, **83**, 2934–8.

Rappold, G. A., Cremer. T., Hager, H. D., Davies, K. E., Muller, C. R. & Yang, T. (1984). Sex chromosome positions in human interphase nuclei as studied by *in situ* hybridization with chromosome specific DNA probes. *Human Genetics*, **67**, 317–25.

Rodeck, C. H. & Campbell, S. (1978). Sampling pure fetal blood by fetoscopy in the second trimester of pregnancy. *British Medical Journal*, **ii**, 728–30.

Saiki, R. K., Gelfand, D. H., Stoffel, S., Scharf, S. J., Higuchi, R., Horn, G. T., Mullis, K. B. & Erlich, H. A. (1988). Primer-directed enzymatic amplification of DNA with a thermostable DNA polymerase. *Science*, **239**, 487–91.

Summers, P. M., Campbell, J. M. & Miller, M. W. (1988). Normal *in-vivo* development of marmoset monkey embryos after trophectoderm biopsy. *Human Reproduction*, **3**, 389–393.

Trask, B., Van den Engh, G., Pinkel, D., Mullikin, J., Waldman, F., Van Dekken, H. & Gray, J. (1988). Fluorescence *in situ* hybridization to interphase cell nuclei in suspension allows flow cytometric analysis of chromosome content and microscopic analysis of nuclear organization. *Human Genetics*, **78**, 251–9.

Varmuza, S., Prideaux, V., Kothery, R. & Rossant, J. (1988). Polytene chromosomes in mouse trophoblast giant cells. *Development*, **102**, 127–34.

Vorsanova, S., Yurov, Y. B., Alexandrov, I. A., Demidova, I. A., Mitkevich, S. P. & Tirskaia, A. F. (1986). 18p- syndrome: An unusual case and diagnosis by *in situ* hybridisation with chromosome 18-specific alphoid DNA sequence. *Human Genetics*, **72**, 185–7.

West, J. D. (1989). The use of DNA probes in preimplantation and prenatal diagnosis. *Molecular Reproduction and Development*, **1**, 138–45.

West, J. D., Gosden, J. R., Angell, R. R., Hastie, N. D., Thatcher, S. S.,

Glasier, A. F. & Baird, D. T. (1987). Sexing the human pre-embryo by DNA–DNA *in situ* hybridisation. *The Lancet*, **i**, 1345–7.

West, J. D., Gosden, C. M., Gosden, J. R., West, K. M., Davidson, Z., Davidson, C. & Nicolaides, K. H. (1989*a*). Sexing the human fetus and identification of polyploid nuclei by DNA–DNA *in situ* hybridisation in interphase nuclei. *Molecular Reproduction and Development*, **1**, 129–37.

West, J. D., Gosden, J. R., Angell, R. R., West, K. M., Glasier, A. F., Thatcher, S. S. & Baird, D. T. (1988). Sexing whole human pre-embryos by *in situ* hybridisation to a Y-chromosome specific DNA probe. *Human Reproduction*, **3**, 1010–19.

West, J. D., West, K. M. & Aitken, R. J. (1989*b*). Detection of Y-bearing spermatozoa by DNA–DNA *in situ* hybridisation. *Molecular Reproduction and Development*, **1**, 201–7.

C. S. Herrington, J. Burns and J. O'D McGee

Non-isotopic *in situ* hybridisation in human pathology

Introduction

In situ hybridisation (ISH) may be defined as the direct detection of nucleic acid in intact cellular material. Nucleic acids may be exogenous or endogenous, nuclear or cytoplasmic, DNA or RNA. A variety of cell and tissue samples can be studied using ISH, from individual chromosomes in metaphase spreads to archival paraffin embedded biopsy material. Using appropriately labelled probes, the presence or absence of normal and abnormal nucleic acids can not only be detected but can also be correlated with cell and tissue morphology. This provides a wealth of information regarding both the genotype and phenotype of cells within pathological lesions and will, by combination with other techniques such as immunocytochemistry, allow greater understanding of the pathophysiology of abnormal cells and the interactions between them.

The technique of *in situ* hybridisation was originally described in 1969 for the detection of abundant ribosomal RNA sequences in non-mammalian systems with ^{32}P-labelled probes (Gall & Pardue, 1969; John *et al.*, 1969). By increasing the sensitivity of detection and resolution of the procedure, using isotopes of high specific activity and shorter track length than ^{32}P (e.g. ^{125}I, ^{35}S, ^{3}H), single copy genes were visualised on chromosomes (Gerhard *et al.*, 1981). In the 1980s, non-isotopic *in situ* hybridisation (NISH) has been developed for gene localisation which is as sensitive as radioisotopic techniques. Single copy genes have now been mapped on chromosomes by NISH (Bhatt *et al.*, 1988). In human disease, ISH can be applied to the detection of normal and abnormal nucleic acids. Development of techniques for ISH has been directed, in the context of laboratory medicine, towards procedures which are clinically useful. This requires not only the production of clinically relevant information but also applicability to routine laboratory testing. The use of radioactive reporter molecules is not suitable for this purpose because detection by autoradiography takes days to weeks for ^{3}H compared to 2–5 hours for non-isotopic reporters. Moreover, radioactive waste disposal is becoming an ecological problem. Non-isotopic probe labels, detected by histochemical means, obviate the need for radiation protection facilities. Non-isotopic procedures are more rapid, provide greater cellular resolution, and allow the use of staining reagents to highlight cellular details of interest in a NISH preparation.

Many different non-isotopic compounds have been used as probe labels (Matthews & Kricka, 1988). The vitamin, biotin, was the first reporter to be introduced. This was employed in NISH to utilise its unique property of high-affinity binding to the naturally occurring protein avidin (Langer et al., 1981). This interaction has a dissociation constant (K_D) of 10^{-15} M compared with, at best, 10^{-12} M for antibody–antigen interactions. In addition, avidin is tetravalent and is therefore capable of binding twice as many molecules of biotin per molecule of protein as antibodies. Many different methods of biotin detection have been developed, utilising avidin and/or antibiotin antibodies conjugated to enzymes or fluorochromes. A disadvantage of biotin, however, is that it is present normally in many tissues, particularly the liver; this produces high background noise and can complicate interpretation. As a result, increasing use is being made of alternative reporters. These are being used in combination with biotin for the detection of different sequences in the same cell, and as a replacement for biotin in the detection of unique sequences. These alternatives include aminoacetylfluorene (AAF) (Landegent et al., 1984), mercury (Hopman et al., 1986) and digoxigenin (Herrington et al., 1989a, b).

The choice of reporter molecule is governed by considerations of routine applicability, safety, sensitivity and the availability of simple detection systems. For example, AAF is potently carcinogenic; mercuric cyanide is toxic. However, as mercury is introduced chemically, it can be used to label nucleic acids in any form. The precise choice must be determined by the facilities available and the nature of the target nucleic acid to be detected.

Having chosen a reporter molecule and acquired the appropriate nucleic acid probe, the probe is labelled. This is done in a variety of ways and, again, depends on the purpose of the investigation. For high sensitivity, it has been advocated that probes labelled by nick translation are more suitable because of the potential for probe networking (Gerhard et al., 1981). Similar considerations apply to random priming but this requires a linear template molecule and cannot be performed on circular plasmid DNA. In our experience, random primed labelling also requires that the labelled probe be decreased in size to 100–400 bp by DNase treatment in order to achieve maximum signal; large probes presumably do not easily permeate intact cells. Oligonucleotide probes in NISH are less sensitive because of their short length. Although, in theory, an oligonucleotide greater than 16 base pairs long is unique in the human genome (Thein & Wallace, 1986), specificity must be determined for individual oligonucleotide probes. Oligonucleotides can, however, be tailor-made for any sequenced nucleic acid. Both sensitivity and specificity can be increased by lengthening the oligonucleotide but cost increases accordingly. Oligonucleotide cocktails have been used to enhance sensitivity but specificity of probe must be ensured for each oligonucleotide individually. As they are single stranded, labelling is performed by either enzymatic addition of labelled nucleotides (biotin, digoxigenin, etc) to the 5' or 3' terminus, by chemical addition; or by photobiotinylation (Chan & McGee, 1989). There is debate about the degree of base substitution which is optimal for *in situ* hybridisation. We have found that the degree of substitution of both biotin and digoxigenin nucleotides has little effect on the signal obtained: the most important parameter is probe size (Evans & McGee, unpublished data).

The target nucleic acid within cells and tissues is surrounded by other cellular constituents and is cross-linked both to other nucleic acids and associated proteins in aldehyde fixed biopsies (the routine fixation method in almost all pathology laboratories). These cross-linked components must be removed to allow access of probe to target nucleic acid but sufficient supporting material must remain to maintain cell and tissue morphology, and to prevent loss of target nucleic acid. The process of unmasking by proteolysis must be carefully controlled and may need to be optimised for each experimental system. The two enzymes most widely used for this purpose are proteinase K (Burns *et al.*, 1988; Syrjanen *et al.*, 1988) and pepsin-HCl (Burns *et al.*, 1987). The former is particularly useful for unmasking protein-coated viral DNA. The latter, however, is less effective for this purpose but is more appropriate for endogenous genomic DNA and some mRNA (Burns *et al.*, 1987; Dirks *et al.*, 1989). The use of aminopropyltriethoxysilane as a section adhesive permits the use of higher concentrations of enzyme without loss of cellular material (Burns *et al.*, 1988). This is particularly important in the study of viral infection of tissues (see below).

After unmasking, probe and target nucleic acids are denatured simultaneously and allowed to hybridise under appropriate conditions. Following hybridisation, preparations are washed in solutions containing combinations of salt and formamide. Mismatched hybrids are removed in this way to a level of stringency determined by the salt/formamide/temperature combination. Prior to detection of the reporter molecule, non-specific binding of antibody and avidin is reduced by a variety of blocking methods. Preincubation with bovine serum albumin reduces the binding of other proteins through electrostatic forces and is widely used for this purpose. However, non-fat dry milk (Duhamel & Johnson, 1985) was found to be superior to bovine serum albumin for the inhibition of non-specific nuclear staining and this has been used successfully by others (Pinkel *et al.*, 1988; Herrington *et al.*, 1989c). High salt concentrations (Singer *et al.*, 1987) and high pH washing conditions have been advocated for the reduction of non-specific avidin binding but high salt is only applicable to single step avidin detection as antigen/antibody interactions are not stable under these conditions. Similarly, high pH washing is not universally applicable (Duhamel & Johnson, 1985).

A variety of detection systems are available for non-isotopic reporter molecules; these include both immunocytochemical and affinity cytochemical techniques. The presence of hybrids between labelled probe and target DNA can be visualised either by fluorescent molecules or by enzyme/chromogenic substrate combinations. Fluorescence techniques allow localisation of signal with high resolution and sensitivity (Lawrence et al., 1988) and are particularly useful for the analysis of single cells. The use of different reporter molecules detected by fluorescent indicators of contrasting colours has been applied to the visualisation of more than one nucleic acid sequence (see below). However, fluorescence microscopy does not allow the simultaneous analysis of cellular morphology and cannot readily be applied to the study of tissue sections because of autofluorescence. In addition, the equipment used for fluorescence microscopy is expensive and, more importantly, most fluorescent labels fade on exposure to light of the appropriate wavelength, even if stabilised using various antifade compounds (Johnson et al., 1982). Experiments cannot therefore be re-examined retrospectively. Thus, for routine use, nonfluorescent detection of nonisotopic reporters is the method of choice. These methods produce stable coloured reaction products which can be analysed using conventional light microscopy.

In all experimental systems, adequate controls must be employed to ensure specificity of signal. The hybridisation of cells and tissues with hybridisation mix alone allows the contribution of the detection system to the final signal to be determined. The use of labelled vector only (into which DNA is inserted) as a probe ensures that signal is not due to vector sequences contained in nick translated or random primed probes. Hybridisation with an irrelevant labelled probe ensures specificity of signal. Positive controls, for example labelled total human DNA (Burns et al., 1986), ensure that all components of the systems are working. Finally, the competitive inhibition of probe binding by the addition of excess unlabelled probe can be used as a specific negative control.

The application of NISH to human disease involves the detection of normal and abnormal nucleic acids. These may be DNA or RNA, nuclear or cytoplasmic, endogenous or exogenous. A variety of cell or tissue samples can be studied, from individual chromosomes in metaphase spreads to archival paraffin-embedded biopsies. Using non-isotopically labelled probes, the presence or absence of normal and abnormal nucleic acids can be correlated with cell and tissue morphology. This approach is summarised in Figure 1.

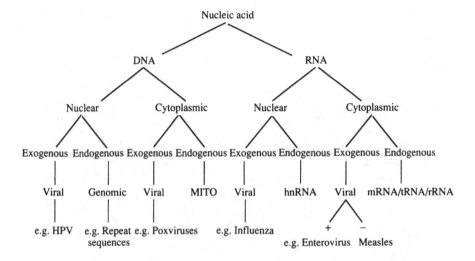

Figure 1. Classification of nucleic acids detectable by *in situ* hybridisation. Mito = mitochondrial DNA; hnRNA = heterogeneous nuclear RNA; mRNA = messenger RNA; tRNA = transfer RNA; rRNA = ribosomal RNA; HPV = human papillomavirus; + refers to positive, i.e. sense strand RNA viruses; − refers to negative or anti-sense strand RNA viruses.

Endogenous nuclear DNA: interphase cytogenetics

Several families of repetitive DNAs are present throughout the genome. These are interspersed with genes present in lower numbers, including single copy genes. Genomic DNA can be studied using metaphase chromosome spreads and interphase cells. The use of metaphase spreads allows precise localisation of hybridisation signal relative to bands produced by standard cytogenetic procedures. Using this method, single copy genes can be detected and mapped (Bhatt *et al.*, 1988) (Fig. 2). Similarly repetitive DNA sequences can be visualised on chromosomes (Burns *et al.*, 1985; Cremer *et al.*, 1986) (Fig. 3); the precise location and the specificity of the repetitive probes can be determined as a prelude to their use in interphase cytogenetics. This also gives direct information about the distribution of repeat sequences in the normal genome. This has yielded information about short interspersed repeated sequences (SINEs) (Houck *et al.*, 1979; Korenberg & Rykowski, 1988), long, interspersed repeated sequences (LINEs) (Shafit-Zagardo *et al.*, 1982; Korenberg & Rykowsky, 1988), centromeric (alphoid) repeats (Cremer *et al.*, 1986; Maio *et al.*, 1981; Devilee *et al.*, 1986) and more recently midisatellite sequences (Nakamura *et al.*, 1987*a*, *b*; Buroker *et al.*, 1987). These repeats have been used in interphase

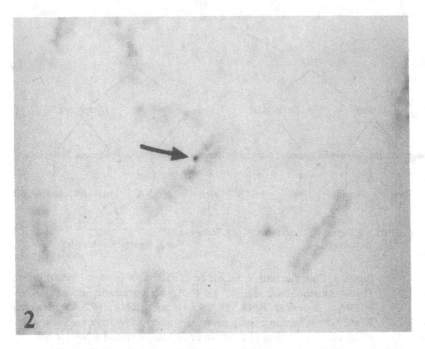

Figure 2. The location of N-ras on the short arm of chromosome 1 is indicated by an arrow on this metaphase spread of normal white blood cells. See Bhatt et al., 1988 for methodology.

cells and have provided information about normal and abnormal interphase nuclei. The logical extension of this approach is the detection of unique sequences in the interphase nucleus, ultimately in archival tissue blocks; present methodology, as yet, precludes this approach.

Using fluorescent detection systems in a cell line containing Epstein Barr virus, the integrated viral genome has been mapped at the single copy level (Lawrence et al., 1988) but the statistics of the cells labelled is omitted from this report. A statistical analysis of male interphase nuclei labelled with biotinylated probes for hypoxanthine-guanine phosphoribosyltransferase (HGPRT) on the X chromosome, α–1-antitrypsin gene on chromosome 14 and N-ras on chromosome 1, has shown distributions of dot number per cell with median numbers of 1 dot for HGPRT and 2 dots for α–1-antitrypsin (Fig. 4); this is in agreement with the expected dot number (Bhatt & McGee, unpublished). The sensitivity of detection systems used at present in archival material is approximately 30 copies (Herrington et al., 1989a; Syrjanen et al., 1988; Burns et al., 1987; Walboomers et al., 1988). Amplification of the detection systems for digoxigenin and biotin labelled probes increases this

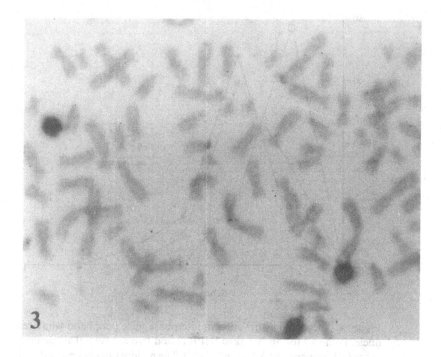

Figure 3. Male lymphocyte metaphase spreads were probed with biotinylated pHY2.1, a probe which is relatively specific for the long arm of the Y chromosome. Three Y chromosomes are labelled black. The probe was detected using anti-biotin peroxidase and silver enhancement (Burns et al., 1985, 1986).

sensitivity at least tenfold and therefore single copy sequences may soon be detectable in archival material (Herrington & McGee, unpublished data).

Exogenous nuclear DNA

The presence of non-human DNA in the nucleus can be determined by using DNA probes specific for the sequence of interest. Exogenous DNA is primarily of viral origin. These may be double-stranded DNA viruses (Baltimore Class I), single-stranded DNA viruses (Baltimore Class II), retroviruses (Baltimore Class VI) and those with a mixed genome (Taussig, 1984). The nucleic acid may remain separate from the host genome (episomal) or may integrate into one or more chromosomes. Both may be detected by *in situ* hybridisation and may be distinguishable by their different signal morphology. Many episomal viruses can be identified using conven-

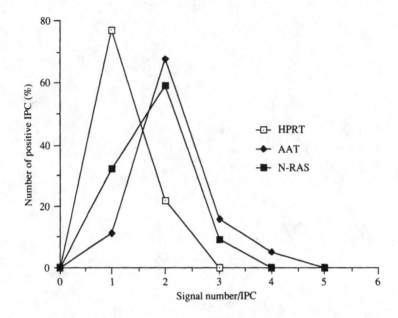

Figure 4. The signal distribution in interphase male peripheral white cell nuclei is shown for an X linked (HPRT) and two autosomal single copy genes (α-1-antitrypsin (AAT) and N-ras). The majority of cells probed with biotinylated HPRT contain one signal while biotinylated AAT and N-ras give two signals. IPC: Interphase cell.

tional stains by the presence of inclusion bodies in the cell nucleus (e.g. cytomegalovirus-CMV) but this is usually of low sensitivity and specificity compared with the detection of nucleic acid (Grody et al., 1987). The most extensively studied by NISH are the human papillomaviruses (HPV) (Burns et al, 1987; Syrjanen, 1987). Episomal viruses are replicating and therefore present in large numbers. In contrast, integrated viruses may be present in very low numbers (Popescu et al., 1987a, b; Mincheva et al., 1987) and therefore require sensitive procedures for their detection. The sensitivity of non-isotopic procedures (see above) presently in routine use in cross-linked archival biopsies is not sufficient to detect viruses integrated in numbers lower than approximately 10–30 per cell (Burns et al., 1987; Syrjanen et al., 1988). However, by increasing the sensitivity of the detection systems, it is likely that low copy number viruses will be detectable in the near future (Herrington & McGee, unpublished data).

The most widely studied of the groups mentioned above are the double-stranded DNA viruses. These comprise in particular the herpes virus and papovavirus groups. Herpes viruses integrate into the host genome (Taussig,

1984) and produce lytic infections of human tissues, particularly at the extremes of age and in immunosuppressed patients. The diagnosis of CMV infection is often difficult, particularly in patients with other clinical problems secondary to underlying disease or its treatment. Conventional diagnostic techniques rely on the regular morphological detection of viral inclusions in cytological or tissue biopsies, or the immunocytochemical detection of viral antigens (Brigati et al., 1983). These procedures are insensitive and therefore the direct detection of viral nucleic acid has been investigated. The CMV genome can be visualised by ISH (Brigati et al., 1983; Wolber & Lloyd, 1988) and this approach has been shown to be of greater sensitivity than immunocytochemistry (Brigati et al., 1983; Wolber & Lloyd, 1988). However, the diagnostic sensitivity of this approach remains to be fully evaluated. Herpes simplex and zoster viruses cause a variety of clinical syndromes. Most diagnoses can be made from clinical features alone but in immunosuppressed individuals the classical clinical symptoms and signs may be masked. This is eminently demonstrated by the dissemination of herpes zoster. The correct diagnosis can be reached in most cases by direct visualisation of the virus by electron microscopy (EM) which is, however, time-consuming. In latent infection, there are no intact virions and EM is not useful; NISH, however, can pick up the viral genome. As herpes viruses integrate in low numbers into the host genome, techniques for their routine direct demonstration by NISH are at present not available. Epstein–Barr Virus (EBV) also produces a variety of clinical syndromes but has been unequivocally associated with human tumours, i.e. Burkitt's lymphoma and nasopharyngeal carcinoma. EBV integrates into the host genome and has been directly demonstrated in interphase nuclei, albeit of cultured cells, by non-isotopic means utilising fluorescent detection methods (Lawrence et al., 1988).

The papovavirus group includes JC virus which is associated with progressive multifocal leucoencephalopathy (PML). The viral genome has been demonstrated by NISH in patients with this disease (Boerman et al., 1989). The human papillomaviruses also belong to this group and are considered in detail below.

Single-stranded DNA viruses are also detectable (Porter et al., 1988) as are viruses with a mixed single- and double-stranded genome such as Hepatitis B virus (Blum et al., 1983).

Endogenous cytoplasmic RNA

RNA is present within the cell nucleus in the form of heterogeneous nuclear RNA (hnRNA). Using fluorescent labelling techniques, nuclear RNA transcripts have been identified within nuclei of cells latently infected with Epstein–Barr virus. This has shown that these transcripts are localised to a particular area of the cell nucleus (Lawrence et al., 1989). Application of these techniques to other specific transcripts will provide a wealth of information on the nuclear anatomy and physiology of transcription. RNA which has undergone post transcriptional modification and is identifiable as mRNA is more directly relevant to the study of human disease. Cytoplasmic RNA can be detected by *in situ* hybridisation using either DNA (Burns et al., 1987) or RNA probes (Mariani-Constantini et al., 1989). cDNA probes are more appropriate in this context than genomic DNA probes as the latter contain intron sequences which have been removed by splicing in the formation of mRNA. More recently, the ability to produce RNA probes (riboprobes) using RNA promoters cloned into plasmids (such as the SP6 series) has allowed the use of these probes to produce more stable RNA–RNA hybrids in the analysis of RNA by ISH.

The study of endogenous RNA by ISH gives information regarding not only the presence of particular RNA species but also quantitative aspects of mRNA production. By enumerating the relative number of silver grains produced by a particular probe in different tissues and comparing with a control probe, the relative production of a particular mRNA species can be estimated, e.g. c-*myc* (Mariani-Constantini et al., 1989). The lack of cellular resolution using radioactive probes is also a problem in the study of mRNA. The track length of ^{3}H is 1 mm in H_2O. It is impossible, therefore, to discriminate signal in closely adjacent cells. This is magnified for ^{35}S which has a much longer track length than ^{3}H. The application of non-isotopic techniques has received less attention for mRNA detection than for genomic DNA. NISH has, however, been used to detect mRNA in cells and tissues (Bresser et al., 1987; Singer et al., 1986). Techniques developed for the detection of multiple nucleic acids require the use of more than one colour and, although differential autoradiography is possible (Hasse et al., 1985a), NISH provides a safer and technically easier approach. Thus, the study of differential expression will require nonisotopic methods.

The ability to detect mRNA in cells has application in several fields. The Reed–Sternberg cells, diagnostic of Hodgkins disease, have been shown by immunocytochemistry to contain immunoglobulins (Garvin et al., 1976). However, immunoglobulin mRNA is not detectable by *in situ* hybridisation. This suggests that the cells do not synthesise immunoglobulin but rather phagocytose it: this has implications for the origin of these cells (Ruprai et al., 1988). The correlation of mRNA synthesis with protein production detected by immunohistochemical techniques allows the cell phenotype to be analysed in more detail. For example, cells in pituitary eosinophil adenomas may produce growth hormone mRNA but not growth hormone itself (Lloyd et al., 1989).

Recently we have been evaluating NISH techniques for mRNA detection in crosslinked archival biopsies. For abundant mRNA species, such as λ and κ Ig chain mRNA present in plasma cells in the lamina propria of the gut, biotinylated oligonucleotides give an excellent signal with clear delineation of cell morphology (Fig. 5).

Exogenous RNA

Non-human RNA can also be detected by ISH and is primarily viral in origin. RNA viruses have been less extensively studied by ISH than DNA viruses but are nevertheless detectable using these techniques (Haase et al., 1985b). The most studied of this group are the measles virus (Haase et al., 1985b) and more recently enteroviruses (Rotbart et al., 1988).

Retroviruses occupy a unique position in the classification presented in Figure 1. Intact virions contain single-stranded RNA which is converted within the host cell to a double-stranded intermediate by reverse transcriptase. This intermediate can then integrate into the host genome and alter host genetic functions. The visna virus, a retrovirus of sheep has been extensively investigated (Haase et al., 1986) and, more recently, ISH technology has been applied to the detection of human immunodeficiency virus (HIV) nucleic acid in patients with HIV related disease (Pezzella et al., 1987).

HPV mRNA, initially rather surprisingly, was visualised in cervical biopsies by NISH (Fig. 6).

Analysis of human disease

In situ hybridisation can be combined with other techniques to provide additional information regarding other aspects of cell or tissue function. In particular, the combination of immunohistochemistry and ISH can be used to demonstrate viral protein production in addition to the viral genome (Fig. 7). By this means, the diagnostic sensitivity of antigen and nucleic acid detection can be compared directly (Table 1). The detection of mRNA in addition to the protein derived by transcription of that RNA allows analysis of cell phenotype in more detail. As mentioned above and discussed in detail below, more than one nucleic acid can be detected within the same cell by NISH. This allows the investigation of the relationship between chromosomes, the prevalence of dual viral infection within individual cells as well as particular lesions, and will be applicable to the study of differential expression.

Having outlined the applications of ISH in the context of human disease, two particular fields will be dealt with in more detail: 1) the diagnosis of viral disease, with particular reference to HPV infection of the uterine cervix, and 2) interphase cytogenetics and the antenatal diagnosis of sex.

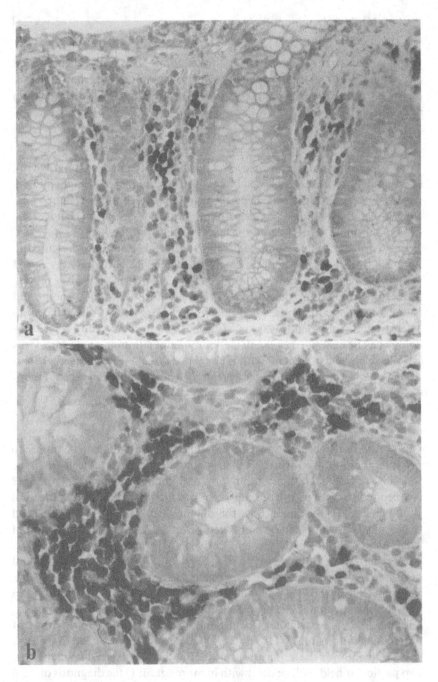

Figure 5. An archival biopsy of human duodenum was probed with a cocktail of biotinylated oligonucleotides complementary to the κ or λ Ig chains. In (a), intense cytoplasmic signal corresponding to λ chains can be seen and in (b) signal due to κ chains.

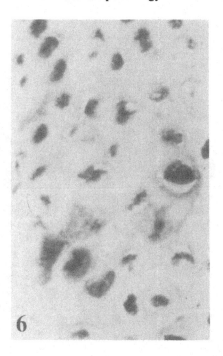

Figure 6. A section from an archival cervical biopsy was probed with biotinylated HPV 6. Both nuclear and cytoplasmic signal can be seen.

The diagnosis of viral disease

It is often difficult to make a precise clinical diagnosis of viral infection. Definite diagnosis is provided by either direct demonstration of the virus, the presence of its characteristic cytopathic effect, or detection of a specific immunological response. The latter, however, depends on the phase of infection and immunocompetence of the host. In addition, widespread exposure of the population to certain types of virus, e.g. Epstein–Barr virus (EBV), can complicate diagnosis as only a specific IgM response is informative. Similarly, anamnestic rises in antibody titre can create diagnostic confusion. Alternatively, the cytopathic effect of viruses can be used as indirect evidence of the presence of a particular viral type, e.g. Negri bodies for rhabdoviruses, Owl's eye inclusions for CMV and koilocytes for HPV. However, the use of morphological criteria for the diagnosis of viral infection is non-specific and insensitive. For example, koilocytes are not indicative of infection by a particular type of HPV and, similarly, not all lesions containing HPVs show koilocytic cells.

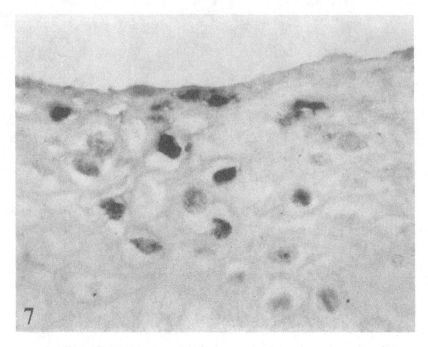

Figure 7. Immunocytochemistry was performed on a section from an archival biopsy using a monoclonal antibody raised to the capsid protein of HPV 16. Signal was developed using a substrate producing a blue colour (seen here as black) and the section then probed with biotinylated HPV 16. The probe detection system produces a red colour (seen here as a darker grey). Superficial cells contain capsid protein (black) and viral DNA but deeper epithelial cells contain only viral genome (darker grey).

Viruses are combinations of nucleic acid and protein. The nucleic acid can be DNA or RNA. The direct demonstration of virus therefore involves the detection of either intact virions, viral proteins or viral nucleic acids. Viral culture provides a definite identification but for some viruses this is either not possible (e.g. HPV) or technically time-consuming. In intact tissues, viruses can be demonstrated directly by immunohistochemistry for viral antigens or by electron microscopy. Immunohistochemistry detects only the antigen to which the antibody is directed and does not visualise virus which is not producing that antigen. CaSki cells, for example, which contain integrated HPV16 sequences (Mincheva et al., 1987), do not express viral capsid protein and therefore do not stain with antibodies to that protein. Electron microscopy is useful for the diagnosis of specific lesions but is only capable of detecting intact virions. Viral subtypes are therefore indistinguishable.

Table 1. *Relative sensitivity of phenotyping and NISH in CIN*

Case No.	HPV Type	Phenotype + ve cells	NISH + ve cells	Ratio (Range)[a]
1	16	165	563	3.4 (2.0–4.5)
2	16	28	160	5.7 (4.5–8.2)
3	16	6	25	4.2 (3.5–4.7)
4	33	160	654	4.1 (3.7–4.3)

[a]The range is derived from counting phenotypically and genotypically positive cells in 3–6 separate areas of each biopsy. These data were generated by analysis of preparations such as those shown in Figure 7.

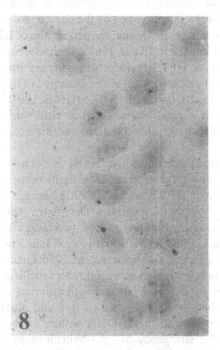

Figure 8. Tissue from a chorionic villus biopsy was probed with biotinylated pHY2.1. Nuclear dots can be clearly seen and indicate that the fetus is male.

Thus, the detection of viral nucleic acid is the only means by which a virologically accurate diagnosis can be made. HPV infection illustrates these principles well. There are over 50 types of HPV, which are distinguished on the basis of nucleic acid sequence (Pfister & Fuchs, 1987). Therefore, typing can only be performed by either DNA extraction and analysis by restriction digestion (De Villiers *et al.*, 1986) or by *in situ* hybridisation (Beckmann *et al.*, 1985). The former 'classic' approach provides sequence data but is time and labour intensive. For diagnostic and investigative purposes, the 'classic' approach is being replaced by the polymerase chain reaction and by NISH (Fig. 6). The latter allows the simultaneous analysis of cell and tissue morphology with the presence and morphology of the specific NISH signal. More recently, HPV infection has been studied by the technique of 'filter *in situ* hybridisation' (Wagner *et al.*, 1984). The latter is a misnomer. 'Filter *in situ* hybridisation' is simply dot hybridisation of cell debris immobilised on a filter matrix.

Anogenital lesions of squamous epithelia have been associated in both sexes with particular HPV types (Syrjanen *et al.*, 1987). There is also a correlation between viral type and the malignant potential of that lesion. Thus, cervical warts and the low grades of cervical epithelial dysplasia (CIN 1/11) are associated with HPV6/11 infection but as the degree of dysplasia increases through CIN 111 to invasive carcinoma, the proportion of HPV16/18/31/33/35 infection increases in parallel (Koss, 1987). Similarly, exophytic penile warts are associated with HPV6/11 infection whereas Bowenoid papulosis, which is premalignant and associated with carcinoma of the cervix in the partners of affected individuals, harbours HPV16 (Gross, 1987). NISH analysis identifies HPV types in clinical biopsies and cervical smears and hence distinguishes lesions which have a low (benign) and high malignant potential.

The detection of more than one nucleic acid in cells and tissues is informative in analysis of putative multiple viral infection of the same tissue. A high percentage of double infection with HPV types 6 and 11 has been reported in several studies of warts from various anatomical sites (Del Mistro *et al.*, 1987; Valejos *et al.*, 1987; Del Mistro *et al.*, 1988). However, two large studies of similar lesions by DNA extraction and restriction digestion failed to find a single case of HPV6/11 double infection (De Villiers *et al.*, 1986; Gissmann *et al.*, 1983). The discrepancy was investigated using a double labelling NISH technique in which the two viral types were labelled with different reporter molecules and were distinguished in different colours (red and blue). This showed that both probes hybridised to target DNA within the same cell nuclei. However, increasing the stringency of post-hybridisation washes removed one signal with retention of the other, suggesting that the dual labelling of cells was due to cross hybridisation of the two probes with one tissue target. This has been confirmed by analysis of extracted DNA by the polymerase chain reaction. The reason for cross-hybridisation in tissue but not on filters despite the use of the same reaction conditions was shown to be due to the elevation of the melting temperatures of hybrids formed in NISH reactions in archival tissue sections (Herrington et al., 1989c).

This double detection procedure has also been applied to the detection of HPVs in cervical smears (Herrington & McGee, unpublished data). By detecting the HPV subtypes associated with benign disease (HPV6/11) with a red substrate and those associated with premalignant and malignant diseases (HPV16/18/31/33/35) with a blue substrate, a single cervical smear can be typed according to its malignant potential. Correlation with cellular morphology allows more detailed investigation of the biology of HPV infection and analysis of the cellular localisation of multiple HPV infections. Thus, double infection of individual cells with both classes of HPV has been identified (Herrington & McGee, unpublished data).

Thus, NISH is a useful means of investigating viral infection of human tissue. Specifically, it provides information on HPV types in different clinical lesions of varying malignant potential, and provides accurate histopathological and molecular biological information on individual cells within a tissue.

Interphase cytogenetics and its applications

Cytogenetics is a hybrid science representing the union of cytology and genetics and, until recently, has been primarily concerned with the chromosomal basis of heredity. This study has involved the analysis of chromosomes during the metaphase of the cell cycle when they are at their most condensed and recognisable as morphologically distinct entities (Swanson, 1957). Until 1970, when banding techniques were introduced, chromosomes were recognisable only on the basis of relative length and the position of the centromere. Banding enables precise identification of chromosomes and structural abnormalities. However, it provides little information about chromosome structure and inter relationships during other phases of the cell cycle.

Individual chromosomes can be visualised in interphase nuclei by both isotopic and non-isotopic *in situ* hybridisation using chromosome specific probes. Isotopic *in situ* hybridisation using probes specific for sequences present on the X and Y chromosomes allowed determination of the position of the sex chromosomes in interphase nuclei (Rappold et al., 1984). The development of non-isotopic techniques and their application to the detection of a Y chromosome specific probe allowed the specific visualisation of a single chromosome in individual cells (Burns et al, 1985). With the use of other chromosome specific probes, this approach has become more widely applicable (Hopman *et al.*, 1989). The term interphase cytogenetics was introduced to describe the analysis of chromosomes in the interphase nucleus using these probes (Cremer *et al.*, 1986) and a variety of such probes has been formulated based on various classes of repetitive DNAs. Many repetitive sequences are dispersed throughout the genome, e.g. SINEs and LINEs (Korenberg & Rykowski, 1988) but some are clustered at either specific centromeres (alphoid repeats) (Devilee *et al.*, 1986) or at less well-defined but chromosome-specific sites; variable number of tandem repeat – VNTR (Nakamura *et al.*, 1987a). Alphoid repeats are relatively (but not absolutely) chromosome specific and have been implicated in chromosome pairing at mitosis. Probes for these regions are chromosome specific at high stringency and are useful for the analysis of numerical chromosome abnormalities. Application of these repeat sequences to the interphase cell nucleus in a variety of preparations allows the numeration of chromosome complement. By using these sequences as probes for ISH both individually and in combination, individual chromosomes and the relationship between them can be visualised in intact cells. To date, this has been achieved only in isolated cell preparations both from cell lines (Cremer *et al.*, 1988a; Devilee *et al.*, 1988; Hopman *et al.*, 1988), or disaggregated tumours (Devilee *et al.*, 1988), using fluorescent detection procedures. A method described recently on model systems (Herrington *et al.*, 1989b) will allow the application of these probes to archival biopsy material so that chromosome number of individual cells within a cancer can be counted.

Alphoid repeats cannot delineate structural chromosome abnormalities which do not involve the centromere. This problem can be approached by either using probes specific for non-centromeric regions in addition to the alphoid repeats or by using chromosome specific libraries, which label the whole chromosome. These libraries were derived from individual flow sorted chromosomes. Hybridisation of metaphase chromosomes with these libraries shows that considerable homology exists between sequences of different chromosomes (Lichter *et al.*, 1988). The addition of unlabelled genomic DNA suppresses the signal from these ubiquitous sequences on other chromosomes: this is termed *in situ* suppression hybridisation (Pinkel *et al.*, 1988; Lichter *et al.*, 1988; Cremer *et al.*, 1988b). The advantage of this approach is that whole chromosomes can be visualised and the origin of structurally abnormal chromosomes can be unequivocally identified. This approach has been used to detect both chromosome translocation and aneuploidy in intact cells. It cannot evaluate less gross abnormalities (Cremer *et al.*, 1988b).

Other chromosome specific probes have also been applied in interphase cytogenetics. Some have been produced by shearing genomic DNA to produce large molecular weight fragments followed by selection of highly repetitive components (Moyzis *et al.*, 1987). More recently, the most divergent part of a chromosome 17 specific clone derived in this way has been used as an oligonucleotide probe. This allows chromosome specificity to be achieved without the use of high stringency conditions (Meyne & Moyzis, 1989). Another group of probes has been derived by screening a human genomic library with oligonucleotides homologous to the pseudo-insulin, zeta globin and myoglobin genes and selecting chromosome specific recombinants. These are known as 'midisatellites' or variable number tandem repeats (VNTRs) (Nakamura et al., 1987b). One of these has been mapped to the telomere of the short arm of chromosome 1 (Buroker et al., 1987) and will thus be useful in combination with centromeric probes and other VNTRs. Indeed, a recent report has demonstrated that centromeric and telomeric probes can be used to label the same chromosome (Van Dekken & Bauman, 1988). At the other end of the spectrum, highly sensitive amplified detection systems have been used in an attempt to detect single copy genes in interphase nuclei. Preliminary results suggest that this is possible using nick translated probes (Herrington & McGee, unpublished data). It is likely that a combination of the methods described above will be required to produce a clear picture of both chromosome structure and abnormality in interphase cells.

A prelude to the study of human disease using these techniques is the analysis of the normal interphase nucleus. This has shown a non-random arrangement of chromosomes, although the three-dimensional (3D) nature of the nucleus projected on to two dimensions produces problems of interpretation (Rappold et al., 1984, Trask et al., 1988). The use of in situ suppression hybridisation in cell lines with stable translocations and known numerical chromosome aberrations has provided insight into the 3D arrangement of abnormal chromosomes (Pinkel et al., 1988).

The application of interphase cytogenetics to human disease can be divided into three broad groups: 1) the detection of normal chromosomes present appropriately in abnormal individuals, e.g. antenatal sex determination in high risk groups for sex linked disease, 2) detection of normal chromosomes present abnormally, e.g. evaluation of ontogeny of cells in patients with sex mismatched transplants, and 3) detection of numerical and structural chromosome abnormalities, e.g. in cancer cells.

1. A repetitive DNA sequence, relatively specific for the Y chromosome (and absolutely so at high stringency) was described by Cooke (Cooke et al., 1982). Although 100–200 copies of highly homologous sequences are present on the autosomes of both males and females, the signal produced by the 2000 copies on the Y chromosome is easily distinguished. In addition, high stringency conditions can be used to render the probe absolutely specific. This probe can be used to identify interphase cells which contain the Y chromosome (Burns et al., 1985). Using non-isotopic detection procedures,

fetal sexing can be performed in under 24 hours on chorionic villus biopsies (Burns *et al.*, 1985): we now do this in less than 4 hours (Fig. 8). This obviates the need for cell culture, and examination of metaphase spreads; this takes about two weeks. The prenatal determination of sex has considerable importance in the genetic counselling of families with X-linked disease. NISH simplifies sex determination and allows decisions to be reached earlier in pregnancy, thus reducing maternal morbidity should termination be deemed necessary (Guyot *et al.*, 1988). Similarly, the sex of embryos can be determined by this method during *in vitro* fertilisation. This has implications for prevention of X-linked disease before implantation (see discussion; Burns *et al.*, 1985).

2. The determination of the origin of cells in patients who have received sex mismatched transplants or grafts has relevance to both the study of the pathophysiology of graft acceptance and to graft evaluation. The analysis of patients who have undergone sex mismatched bone marrow transplants (BMT) has shown that circulating WBCs of dendritic type are derived almost exclusively from the donor (Reittie *et al*, 1988). Thus, the reappearance of recipient cells can be detected early and therapeutic intervention instituted if required. Similarly, the origin of cells comprising tumours arising in patients who have received BMT can be determined: a lymphoma was shown to be of donor origin using the Y chromosome probe (Lyttelton *et al.*, 1988). The study of allografts in patients with burns has shown that cells from the graft are not demonstrable in biopsies taken from the graft area one week after application. Thus allografts appear to act simply as biological dressings and do not survive (Burt *et al.*, 1989; Brain *et al.*, 1989).

3. Chromosome abnormalities are of two main types: structural and numerical. Some well-defined chromosome abnormalities are associated with particular clinical syndromes. The antenatal diagnosis of these has relied on the demonstration of the abnormality in metaphase spreads from cultured fetal cells. Interphase cytogenetics can be used to demonstrate these abnormalities without culture and has been used successfully in the diagnosis of trisomy 21 (Julien *et al.*, 1986). However, the efficiency with which numerical abnormalities can be detected in a small number of cells is unknown as not all cells known to contain a particular number of chromosomes contain the appropriate number of signals (Cremer *et al.*, 1986).

Chromosome abnormalities have been recognised in malignant cells since 1914 (Boveri, 1914) but it is only recently that it has been shown that these abnormalities are non-random. The analysis of the karyotype of malignant cells has until recently been performed using disaggregated tumour cells with or without limited cell culture and colchicine block (Rodgers *et al.*, 1984; Hill *et al.*, 1987). This produces a highly selected population of cells, both in terms of those in the cell cycle at the time of tumour excision and those most rapidly dividing in vitro. The use of chromosome specific probes visualises chromosomes, or parts of chromosomes, in interphase nuclei. Breast (Devilee *et al.*, 1988), bladder (Hopman *et al.*, 1988), gynaecological (Nederlof *et al.*, 1989) and cerebral (Cremer *et al.*, 1988a) tumours have been studied in this way. Most studies have been performed with either cell lines or disaggregated tumours using non-isotopic *in situ* hybridisation and the signal generated by fluorescent means. Although this gives signal of high resolution, its application to tissue sections is not possible. The use of more than one reporter molecule allows the detection of several different chromosomes in individual nuclei, each distinguished in a different colour by means of differential fluorescence. Two, and more recently three, different chromosomes (Nederlof *et al.*, 1989) have been demonstrated in individual nuclei. Non-fluorescent detection of more than one nucleic acid has been achieved (Herrington *et al.*, 1989b) in archival biopsy samples and will be applicable to interphase cytogenetics.

Gross structural chromosome abnormalities such as large deletions and translocation are detectable using *in situ* suppression hybridisation. This has the advantage over conventional cytogenetics that marker chromosomes can be identified.

Thus, interphase cytogenetics is a powerful new tool in genetic analysis, particularly of human tumours and in antenatal diagnosis. Techniques available at present allow the numeration and investigation of structural abnormalities of chromosomes in isolated cells. Application of this to archival biopsies will provide a wealth of cytogenetic information which can be correlated with both histopathological abnormalities and clinical parameters. This may provide useful prognostic markers of therapeutic value in patient management.

Conclusion

Non-isotopic *in situ* hybridisation may be used to study any nucleic acid for which a cloned sequence is available or whose nucleotide sequence is known. Techniques which are routinely applicable to routine biopsy samples are capable of detecting only sequences which are present in numbers of approximately 10–30 or more. Application of repeat sequences has produced useful information of both diagnostic and research importance. Present methodology has not reached the level of sensitivity required to visualise unique human or viral sequences routinely in interphase nuclei. With the rapid advances being made in the field of non-isotopic *in situ* hybridisation, however, it is likely that single copy genes will soon be reliably detected in isolated cells and, ultimately, in archival material. This will open up many areas of investigation. Our main aim is to define the cancer genome *in situ* at the single gene level.

Acknowledgements

We thank Miss Lesley Watts for typing the manuscript. The original work reported was funded by grants from the Cancer Research Campaign to J.O'D.McG. C.S.H. is a CRC Clinical Research Fellow.

References

Beckmann, A. M., Myerson, D., Daling, J. R., Kiviat, N. B., Fenoglio, C. M. & McDougall, J. K. (1985). Detection and localisation of human papillomavirus DNA in human genital condylomas by in situ hybridisation with biotinylated probes. *Journal of Medical Virology*, **16**, 265-73.

Bhatt, B., Burns, J., Flannery, D., McGee, J. O'D. (1988). Direct visualisation of single copy genes on banded metaphase chromosomes by nonisotopic in situ hybridisation. *Nucleic Acids Research*, **16**, 3951-61.

Blum, H. E., Stowring, L., Figus, A., Montgomery, C. K., Haase, A. T., Vyas, G. N. (1983). Detection of hepatitis B virus DNA in hepatocytes, bile duct epithelium and vascular elements by in situ hybridisation. *Proceedings of the National Academy of Sciences, USA*, **80**, 6685-8.

Boerman, R. H., Arnolous, E. P. J., Raap, A. K., Peters, A. C. B., Ter Schegget, J., Van Der Ploeg, M. (1989). Diagnosis of progressive multifocal leucoencephalopathy by hybridisation techniques. *Journal of Clinical Pathology*, **42**, 153-61.

Boveri, T. (1914). *Zur Frage der Entstehung maligner Tumoren*, Fischer, Jenna.

Bresser, J. & Evinger-Hodges, M. J. (1987). Comparison and optimisation of *in situ* hybridisation procedures yielding rapid, sensitive mRNA detections, *Gene Analysis & Technology*, **4**, 89-104.

Brigati, D. J., Myerson, D., Leary, J. J., Spalholz, B., Travis, S. Z., Fong, C. K., Hsiung, G. D. & Ward, D. C. (1983). Detection of viral genomes in cultured cells and paraffin-embedded tissue using biotin-labelled hybridisation probes. *Virology*, **126**, 32-50.

Burns, J., Chan, V. T.-W., Jonasson, J. A., Fleming, K. A., Taylor, S. & McGee, J. O'D. (1985). Sensitive system for visualising biotinylated DNA probes hybridised in situ: rapid sex determination in intact cells. *Journal of Clinical Pathology*, **38**, 1085-92.

Burns, J., Redfern, D. R. M., Esiri, M. M. & McGee, J. O'D. (1986). Human and viral gene detection in routine paraffin embedded tissue by in situ hybridisation with biotinylated probes: viral localisation in herpes encephalitis. *Journal of Clinical Pathology*, **39**, 1066-73.

Burns, J., Graham, A. K., Frank, C., Fleming, K. A., Evans, M. F. & McGee, J. O'D. (1987). Detection of low copy human papilloma virus DNA and mRNA in routine paraffin sections of cervix by non-isotopic *in situ* hybridisation. *Journal of Clinical Pathology*, **40**, 858-64.

Burns, J., Graham, A. K. & McGee, J. O'D. (1988). Non-isotopic detection of in situ nucleic acid in cervix: an updated protocol. *Journal of Clinical Pathology*, **41**, 897–9.

Buroker, N., Bestwick, R., Haight, G., Magenis, R. E. & Litt, M. (1987). A hypervariable repeated sequence on human chromosome 1p36. *Human Genetics*, **77**, 175–81.

Burt, A. M., Pallett, C. D., Sloane, J. P., O'Hare, M. J., Schafler, K. F., Yardeni, P., Eldad, A., Clarke, J. A. & Gusterson, B. A. (1989). Survival of cultured allografts in patients with burns assessed with probe specific for Y chromosome. *British Medical Journal*, **298**, 915–17.

Brain, A., Purkis, P., Coates, P., Hackett, M., Navsaria, H. & Leigh, I. (1989). Survival of cultured allogeneic keratinocytes transplanted to deep dermal bed assessed with probe specific for Y chromosome. *British Medical Journal*, **298**, 917–19.

Chan, V. T.-W. & McGee, J. O'D. (1989). In situ *Hybridisation: Principles and Practice* (eds Polak, J. & McGee. J. O'D.). Oxford: Oxford University Press, in press.

Cooke, H. J., Schmidtke, J. & Gosden, J. R. (1982). Characterisation of a human Y chromosome repeated sequence and related sequence in higher primates. *Chromosoma (Berl.)*, **87**, 491–502.

Cremer, T., Landegent, J., Bruckner, A., Scholl, H. P., Schardin, M., Hager, H. D., Devilee, P., Pearson, P. & van der Ploeg, M. (1986). Detection of chromosome aberrations in the human interphase nucleus by visualisation of specific target DNAs with radioactive and non-radioactive in situ hybridisation techniques: diagnosis of trisomy 18 with probe L1.84. *Human Genetics*, **74**, 346–52.

Cremer, T., Tesin, D., Hopman, A. H. N. & Manuelidis, L. (1988a). Rapid interphase and metaphase assessment of specific chromosomal changes in neuroectodermal cells by in situ hybridisation with chemically modified DNA probes. *Experimental Cell Research*, **176**, 199–220.

Cremer, T., Lichter, P., Borden, J., Ward, D. C. & Manuelidis, L. (1988b). Detection of chromosome aberrations in metaphase and interphase tumour cells by in situ hybridisation using chromosome specific library probes. *Human Genetics*, **80**, 235–46.

De Villiers, E.-M., Schneider, A., Gross, G., & Zur Hausen, H. (1986). Analysis of benign and malignant urogenital tumours for human papillomavirus infection by labelling cellular DNA. *Medical Microbiology and Immunology*, **174**, 281–6.

Del Mistro, A., Braunstein, J. D., Hawler, M. & Koss, L. G. (1987). Identification of human papillomavirus types in male urethral condylomata acuminata by in situ hybridisation. *Human Pathology*, **18**, 936–40.

Del Mistro, A., Koss, L. G., Braunstein, J., Bennett, B., Saccamano, G. & Simons, K. M. (1988). Condylomata acuminata of the urinary bladder: natural history, viral typing and DNA content. *American Journal of Surgical Pathology*, **12**, 205–15.

Devilee, P., Slagboom, P., Cornelisse, C. J., Pearson, P. L. (1986). Sequence heterogeneity within the human alphoid repetitive DNA family. *Nucleic Acids Research*, **14**, 2059–72.

Devilee, P., Thierry, R. F., Kievits, T., Kolluri, R., Hopman, A. H. N., Willard, H. F., Pearson, P. L., Cornelisse, C. J. (1988). Detection of chromosomes aneuploidy in interphase nuclei from human primary breast tumours using chromosome-specific repetitive DNA probes. *Cancer Research*, **48**, 5825–30.

Dirks, R. W., Raap, A. K., Van Minnen, J., Vreugdenhil, E., Smit, A. B. & Van Der Ploeg, M. (1989). Detection of mRNA molecules coding for neuropeptide hormones of the pond snail Lymnaea stagnalis by radioactive and non radioactive *in situ* hybridisation: a model study for mRNA detection. *Journal of Histochemistry & Cytochemistry*, **37**, 7–14.

Duhamel, R. C. & Johnson, D. A. (1985). Use of Nonfat dry milk to block nonspecific nuclear and membrane staining by avidin conjugates. *Journal of Histochemistry & Cytochemistry*, **33**, 711–14.

Gall, J. G. & Pardue, M. L. (1969). Formation and detection of RNA–DNA hybrid molecules in cytological preparations. *Proceedings of the National Academy of Sciences, USA*, **63**, 378–83.

Garvin, A. J., Spicer, S. S. & McKeever, P. E. (1976). The cytochemical demonstration of intracellular immunoglobulin in neoplasms of lymphoreticular tissue. *American Journal of Pathology*, **82**, 457–70.

Gerhard, D. D., Kawasaki, E. S., Bancroft, F. C. & Szabo P. (1981). Localisation of a unique gene by direct hybridisation *in situ*. *Proceedings of the National Academy of Sciences, USA*, **78**, 3755–9.

Gissmann, L., Wolnik, L., Ikenberg, H., Kolclovsky, U., Schnurch, H. G., Zur Hausen, H. (1983). Human Papillomavirus Types 6 and 11 DNA sequences in genital and laryngeal papillomas in some cervical smears. *Proceedings of the National Academy of Sciences, USA*, **80**, 560–3.

Grody, W. W., Cheng, L. & Lewin, K. J. (1987). In situ viral DNA hybridisation in diagnostic surgical pathology. *Human Pathology*, **18**, 535–43.

Gross, G. (1987), In *Papillomoviruses and Human Disease* (eds Syrjanen, K., Gissmann, L. & Koss, L. G.) pp. 197–234. Berlin, New York, London: Springer Verlag.

Guyot, B., Bazin, A., Sole, Y., Julien, C., Daffos, F & Forestier F. (1988). Prenatal diagnosis with biotinylated chromosome specific probes. *Prenatal Diagnosis*, **8**, 485–93.

Haase, A. T., Walker, D., Ventura, P., Geballe, A., Blum, H., Brahic, M., Goldberb, R. & O'Brien, K. (1985*a*). Detection of two viral genomes in single cells by double-label hybridisation *in situ* and color microradioautography. *Science*, **227**, 189–92.

Haase, A. T., Gantz, D., Eble, B., Walker, D., Stowring, L., Ventura, P., Blum, H., Wietgrefe, S., Zupancic, M., Tourtellotte, W., Gibbs, C. J., Norrby, E. & Rozenblatt, S. (1985*b*). Natural history of restricted synthesis and expression of measles virus genes in subacute sclerosing panencephalitis. *Proceedings of the National Academy of Sciences, USA*, **82**, 3020–4.

Haase, A. T. (1986). Analysis of viral infections by *in situ* hybridisation. *Journal of Histochemistry & Cytochemistry*, **34**, 27–32.

Herrington, C. S., Burns, J., Graham, A. K., Evans, M. F., McGee, J. O'D. (1989a). Interphase cytogenetics using biotin and digoxigenin labelled probes I: relative sensitivity of both reporters for detection of HPV16 in CaSki cells. *Journal of Clinical Pathology*, **42**, 592–600.

Herrington, C. S., Burns, J., Graham, A. K., Bhatt, B. & McGee, J. O'D. (1989b). Interphase cytogenetics using biotin and digoxigenin labelled probes II: simultaneous detection of two nucleic acid species in individual nuclei. *Journal of Clinical Pathology*, **42**, 601–6.

Herrington, C. S., Flannery, D. & McGee, J. O'D. (1989). Single and simultaneous nucleic acid detection in archival human biopsies: application of non isotopic *in situ* hybridisation (NISH) and the polymerase chain reaction (PCR) to the analysis of human and viral genes. In (eds Polak, J. & McGee, J. O'D.) In situ *Hybridisation: Principles and Practice*. Oxford: Oxford University Press, in press.

Hill, S. M., Rodgers, C. S. & Hulten, M. A. (1987). Cytogenetic analysis in human breast carcinoma. II. Seven cases in the triploid/tetraploid range investigated using direct preparations. *Cancer Genetics & Cytogenetics*, **24**, 45–62.

Hopman, A. H. N., Wiegant, J. Tesser, G. I. & Van Duijn, P. (1986). A non-radioactive *in situ* hybridisation method based on mercurated nucleic acid probes and sulfhydryl-hapten ligands. *Nucleic Acids Research*, **14**, 6471–88.

Hopman, A. H. N., Ramaekers, F. C. S., Raap, A. K., Beck, J. L. M., Devilee, P., Van Der Ploeg, M. & Vooijs, G. P. (1988). *In situ* hybridisation as a tool to study numerical chromosome aberrations in solid bladder tumours. *Histochemistry*, **89**, 307–16.

Hopman, A. H. N., Ramaekers, F. C. S. & Vooijs, E P. (1990) (April). In In situ *Hybridisation: Principles and Practice* (eds Polak, J. & McGee, J. O'D), Oxford: Oxford University Press, in press.

Houck, C. M., Rinehart, F. P. & Schmid, C. W. (1979). A ubiquitous family of repeated DNA sequences in the human genome. *Journal of Molecular Biology*, **132**, 289–306.

John, H. A., Birnstiel, M. L. & Jones, K. W. (1969). RNA–DNA hybrids at the cytological level. *Nature (London)*, **233**, 582–627.

Julien, C., Bazin, A., Guyot, B., Forestier, F. & Daffos, F. (1986). Rapid prenatal diagnosis of Down's syndrome with *in situ* hybridisation of fluorescent DNA probes. *Lancet* **ii**, 863–4.

Korenberg, J. R. & Rykowski, M. C. (1988). Human genome organisation: Alu, Lines, and the molecular structure of metaphase chromosome bands. *Cell*, **53**, 391–400.

Koss, L. G. (1987). *Papillomaviruses and Human Disease* (eds Syrjanan, K., Gissmann, L. & Koss, L. G.) pp. 235–67. Berlin, New York, London:

Springer-Verlag.
Landegent, J. E., Jansen in de Wal, N., Baan, R. A., Hoeijmakers, J. H. J. & Van der Ploeg, M. (1984). 2-acetylaminofluorene-modified probes for the indirect hybridocytochemical detection of specific nucleic acid sequences. *Experimental Cell Research*, **153**, 61–72.
Langer, P. R., Waldrop, A. A. & Ward, D. C. (1981). Enzymatic synthesis of biotin-labelled polynucleotides: novel nucleic acid affinity probes, *Proceedings of the National Academy of Sciences, USA*, **78**, 6633–7.
Lawrence, J. B., Villnave, C. A. & Singer, R. H. (1988). Sensitive, high-resolution chromatin and chromosome mapping *in situ*: presence and orientation of two closely integrated copies of EBV in a lymphoma line. *Cell*, **52**, 51–61.
Lawrence, J. B., Singer, R. H. & Marselle, L. M. (1989). Highly localised tracks of specific transcripts within interphase nuclei visualised by *in situ* hybridisation. *Cell*, **57**, 493–502.
Lichter, P., Cremer, T., Borden, J., Manuelidis, L. & Ward, D. C. (1988). Delineation of individual human chromosomes in metaphase and interphase cells by in situ suppression hybridisation using recombinant DNA libraries. *Human Genetics*, **80**, 224–34.
Lloyd, R. V., Cano, M., Chandler, W. F., Barkan, A. L., Horvath, E. & Kovaks, K. (1989). Human growth hormone and prolactin secreting pituitary adenomas analysed by *in situ* hybridisation. *American Journal of Pathology*, **134**, 615–13.
Lyttleton, M. P. A., Browett, P. J., Brenner, M. K., Cordingley, F. T. Kohlman, J., McGee, J. O'D., Hamilton-Dutoit, S., Prentice, H. G., Hoffbrand, A. V. (1988). Prolonged remission of Epstein–Barr virus associated lymphoma secondary to T cell depleted bone marrow transplantation. *Bone Marrow Transplantation*, **3**, 641–6.
Maio, J. J., Brown, F. L. & Musich, P. R. (1981). Toward a molecular paleontology of primate genomes 1: the Hind111 and EcoR1 dimer families of alphoid DNAs. *Chromsoma (Berl.)*, **83**, 103–25.
Mariani-Constantini, R., Theillet, C., Hutzell, P., Merlo, G., Schlom, J. & Callahan, R. (1989). *In situ* detection of c-myc mRNA in adenocarcinomas, adenomas, and mucosa of human colon. *Journal of Histochemistry & Cytochemistry*, **37**, 293–8.
Matthews, J. A. & Kricka, L. J. (1988). Analytical strategies for the use of DNA probes. *Analytical Biochemistry*, **169**, 1–25.
Meyne, J. & Moyiz, R K. (1989). Human chromosome-specific repetitive DNA probes: targeting *in situ* hybridisation to chromosome 17 with a 42-base pair alphoid DNA oligomer. *Genomics*, **4**, 472–8.
Mincheva, A., Gissmann, L. & Zur Hausen, H. (1987). Chromosomal integration sites of human papillomavirus DNA in three cervical cancer cell lines mapped by *in situ* hybridisation. *Medical Microbiology & Immunology*, **176**, 245–56.
Moyzis, R. K., Albright, K. L., Bartholdi, L. S., Cram, L. L., Deaven, C. E., Hildebrand, N. E., Joste, J. L., Beyne, J. & Schwarzacher-Robinson, T.

(1987). Human chromosome-specific repetitive DNA sequences: Novel markers for genetic analysis. *Chromosome (Berl.)*, **95**, 375–86.

Nakamura, Y., Leppert, M., O'Connell, P., Wolff, R., Holm, T., Culver, M., Martin, C., Fujimoto, E., Hoff, M., Kumlin, E. & White, R. (1987a). Variable number of tandem repeat (VNTR) markers for human gene mapping. *Science*, **235**, 1616–22.

Nakamura, Y., Julier, C., Wolff, R., Holm, T., O'Connell, P., Leppert, M., & White, R. (1987b). Characterisation of a human 'midisatellite' sequence. *Nucleic Acids Research*, **15**, 2537–46.

Nederlof, P. M., Robinson, D., Abuknesha, R., Wiegant, J., Hopman, A. H. N., Tanke, H. J. & Raap, A. K. (1989). Three-color fluorescence *in situ* hybridisation for the simultaneous detection of multiple nucleic acid sequences. *Cytometry*, **10**, 20–7.

Pezzella, M., Pezzella, F., Galli, C., Macchi, B., Verani, P., Sorice, F. & Baroni, C. D. (1987). *In situ* hybridisation of human immunodeficiency virus (HTLV-III) in cryostat sections of lymph nodes of lymphadenopathy syndrome patients. *Journal of Medical Virology*, **22**, 135–42.

Pfister, H. & Fuchs, P. G. (1987), In *Papillomaviruses and Human Disease* (eds Syrjanen, K., Gissmann, L. & Koss, L. G.) pp. 1–18. Berlin, New York, London: Springer-Verlag (1988).

Pinkel, D., Landegent, J., Collins, C., Fuscoe, J., Segraves, R., Lucas, J & Gray, J. (1988). Fluorescence *in situ* hybridisation with human chromosome specific libraries: Detection of trisomy 21 and translocations of chromosome 4. *Proceedings of the National Academy of Sciences, USA*, **85**, 9138–47.

Popescu, N. C., Amsbaugh, S. C. & DiPaolo, J. A. (1987a) HPV type 18 DNA is integrated at a single chromosome site in cervical carcinoma cell line SW756. *Journal of Virology*, **51**, 1682–5.

Popescu, N. C., DiPaolo, J. A. & Amsbaugh, S. C. (1987b). Integration sites of HPV18 DNA sequences on HeLa cell chromosomes. *Cytogenetics & Cell Genetics*, **44**, 58–62.

Porter, H., Quantrill, A. M. & Fleming, K. A. (1988). B19 parvovirus infection of myocardial cells. *Lancet* i, 535–6.

Rappold, G. A., Cremer, T., Hager, H. D., Davies, K. E., Muller, C. R., Yang, T. (1984). Sex chromosome positions in human interphase nuclei as studied by *in situ* hybridisation with chromosome specific DNA probes. *Human Genetics*, **67**, 317–25.

Rodgers, C. S., Hill, S. M. & Hulten, M. A. (1984). Cytogenetic analysis in human breast carcinoma. I. Nine cases in the diploid range investigated using direct preparations. *Cancer Genetics & Cytogenetics*, **13**, 95–119.

Rotbart, H. A., Abzug, M. J., Murray, R. S., Murphy, N. L. & Levin, M. J. (1988). Intracellular detection of sense and anti-sense enteroviral RNA by *in situ* hybridisation. *Journal of Virological Methods*, **22**, 295–301.

Ruprai, A. K., Pringle, J. H., Angel, C., Kind C. L. & Lauder, I. (1988). Localisation of immunoglobulin light chain mRNA expression in Hodgkins disease by *in situ* hybridisation using biotinylated oligonucleotide probes. *Journal of Pathology*, **154**, 38A.

Reittie, J. E., Poulter, J. W., Prentice, H. G., Burns, J., Drexler, H. G., Balfour, B., Clarke, J., Hoffbrand, A. V., McGee, J.O'D. & Brenner, M. K. (1988). Differential recovery of phenotypically and functionally distinct circulating antigen-presenting cells after allogenic bone marrow transplantation. *Transplantation*, **45**, 1084–91.

Shafit-Zagardo, B., Maio, J. J. & Brown, F. L. (1982). Kpnl families of long interspersed repetitive DNAs in human and other primate groups. *Nucleic Acids Research*, **10**, 3175–93.

Singer, R. H., Lawrence, J. B. & Villnave, C. (1986). Optimisation of in situ hybridisation using isotopic and nonisotopic detection methods. *Biotechniques*, **4**, 230–50.

Singer, R. H., Lawrence, J. B. & Rashtchian, R. N. (1987). In In situ *Hybridisation: Applications to Neurobiology* (eds Valentino, K. L., Eberwine, J. H. & Barchas, J. D.) pp. 71–96, Oxford: Oxford University Press.

Swanson, C. P. (1957). *Cytology and Cytogenetics.* London, New York: Macmillan and Co Ltd.

Syrjanen, K. J. (1987). Biology of HPV infections and their role in squamous cell carcinogenesis. *Medical Biology*, **65**, 21–39.

Syrjanen, S., Partanen, P., Mantyjarvi, R. & Syrjanen, K. (1988). Sensitivity of in situ hybridisation techniques using biotin and ^{35}S labelled human papillomavirus (HPV) DNA probes. *Journal of Virological Methods*, **19**, 225–38.

Taussig, M. J. (1984). *Processes in Pathology and Microbiology*: Oxford, London & Edinburgh: Blackwell Scientific.

Thein, S. L. & Wallace, R. B. (1986). In (ed. Davies, K. E.) *Human Genetic Diseases: A Practical Approach*), pp. 33–50. Oxford: IRL Press.

Trask, B., van den Eugh, G., Pinkel, D., Mullikin, J., Waldman, F., van Dekken, H. & Gray, J. (1988). Fluorescence *in situ* hybridisation to interphase nuclei in suspension allows flow cytometric analysis of chromosome content and microscopic analysis of nuclear organisation. *Human Genetics*, **78**, 251–9.

Valejos, H., Del Mistro, A., Kleinhaus, S., Braunstein, J. D., Halwer, M. & Koss, L. G. (1987). Characterisation of human papilloma virus types in condylomata acuminata in children by *in situ* hybridisation. *Laboratory Investigation*, **56**, 611–15.

Van Dekken, H. & Bauman, J. G. J. (1988). A new application of *in situ* hybridisation: detection of numerical and structural chromosome aberration with a combination centromeric–telomeric DNA probe. *Cytogenetics & Cell Genetics*, **48**, 188–9.

Wagner, D., Ikenberg, H., Boehm, N. & Gissmann, L. (1984). Identification of human papillomavirus in cervical swabs by deoxyribonucleic acid *in situ* hybridisation. *Obstetrics and Gynaecology*, **64**, 767–72.

Walboomers, J. M. M., Melchers, W. J. G., Mullink, H., Meijer, C. J. L. M., Struyk, A., Quint, W. G., Van der Noorda, J. & Ter Schegget, J. (1988).

Sensitivity of *in situ* detection with biotinylated probes of human papolloma virus type 16 DNA in frozen tissue sections of squamous cell carcinomas of the cervix. *American Journal of Pathology*, **131**, 587–94.

Wolber, R. A. & Lloyd, R. V. (1988). Cytomegalovirus detection by nonisotopic *in situ* DNA hybridisation and viral antigen immunostaining using a two-colour technique. *Human Pathology*, **19**, 736–41.

M. Wells

The demonstration of viral DNA in human tissues by *in situ* DNA hybridisation

Introduction

The extraction of single-stranded human DNA, its separation by gel electrophoresis, its transfer to a nitrocellulose filter (the Southern blot) followed by hybridisation with a radioactively labelled viral probe consisting of a complementary DNA sequence forms the fundamental basis of the now well-established technique of DNA hybridisation. Dot blot hybridisation is a more rapid variation of this basic technique.

However, the histopathologist is concerned with morphology and topographical relationships within tissue sections and the investigation of DNA bands does not allow one to precisely localise viral DNA within human tissues. But now the methods of molecular biology and immunohistochemistry have come together to allow the specific recognition of viral DNA in tissue sections.

Since 1985 there has been an increasing number of publications on the subject of *in situ* hybridisation for human papillomavirus (HPV) DNA. Our own work in this field was always aimed at developing a technique that was applicable to the routinely fixed and processed material in a hospital diagnostic histopathology laboratory (Lewis *et al.*, 1987; Wells *et al.*, 1987). Such a technique would permit the retrospective evaluation of archival material, thus considerably widening the scope for investigating the viral status of a variety of pathological lesions.

Numerous detection systems have been developed to visualise labelled gene probes. Autoradiography has been used to detect radio-labelled probes in cell smears and paraffin sections (Gupta *et al.*, 1985). Although autoradiographical techniques are very sensitive, the use of radio-labelled material is unsuitable for most routine applications. The introduction of biotinylated DNA probes (Langer, Waldrop & Ward 1981; Singer & Ward 1982) and the development of methods applicable to formalin-fixed paraffin material has enabled these techniques to become almost routine. These methods involve the use of an anti-biotin antibody subsequently visualised by a fluorescein or enzyme-labelled second antibody. Alternatively, techniques have been devised which utilise the high affinity between biotin (a small water soluble vitamin MW = 228) and the protein tetramer streptavidin (MW = 60000) which contains four biotin binding sites. Horseradish peroxidase has usually been employed as the enzyme label of choice in these detection systems, the diamino benzidine reaction product usually being enhanced by a silver precipitation reaction (Gallyas, Görcs & Merchenthaler 1982).

Lewis et al. (1987) developed a method for the detection of in situ hybridised, biotinylated DNA probes using a streptavidin/polyalkaline phosphatase complex which matched the sensitivities claimed for silver-enhanced perodixase methods.

Theoretically, in situ hybridisation may consist of DNA hybridising to DNA, RNA to DNA or RNA to RNA. Our own technique to date has only employed DNA–DNA hybridisation.

There are several important considerations to ensure successful in situ DNA hybridisation (Crum et al., 1988). The first of these is the proper treatment of slides to maximise adherence of tissue during the procedure. In our early studies we utilised poly-L-lysine coated slides, but there was quite a high rate of tissue loss from the slides. There are other disadvantages of poly-L-lysine; in excess it forms smears which take up many routine stains, the coating encourages bacterial growth on the slide and it is expensive. We now use slides coated with 3-aminopropyltriethoxysilane which has proved to be very effective (Burns et al., 1987). Appropriate tissue fixation is important though our aim was always to develop a technique that was applicable to formalin fixed tissue.

It has been shown that positive results can be obtained with a variety of different fixatives and it may be that modified Carnoy's fixative and periodate–lysine–paraformaldehyde produce better results than formalin-fixed tissue (McAllister & Rock, 1985). The use of mercury based fixatives such as formal in sublimate have not been adequately investigated but DNA extraction studies from routinely processed tissue blocks suggest that they may result in unacceptable degradation of DNA.

Pretreatment of the tissue is essential to ensure that the probe penetrates maximally. Also to be considered are the specific conditions of pre-hybridisation, hybridisation and post-hybridisation washes which are all discussed in more detail below.

Human papillomaviruses

The occurrence of sexually transmitted anogenital warts or condylomata has been recognised since classical times though a viral aetiology was only suggested in the 1920s. In the late 1950s it was first suggested that 'koilocytosis' was a manifestation of viral infection.

Koilocytosis is characterised by cytoplasmic vacuolation and nuclear pyknosis within squamous epithelium. In the late 1960s electron microscopy first showed viral particles within a condyloma. However, it was not until 1976 that condylomatous lesions of the cervix as well as of the external genitalia were recognised. There was also, around this time, a gradual realisation that these condylomata could rarely undergo malignant transformation. In the early 1980s it was realised that human papillomavirus could also cause so-called non-condylomatous infection of the cervix or 'flat warts'.

Throughout the 1980s the increasing association between human papillomavirus infection and squamous neoplasia of the lower female genital tract has emerged.

The tools of electron microscopy and immunohistochemistry have been used to demonstrate viral particles or viral antigen with limited success (Sato et al., 1987; Wilbur, Reichman & Stoler, 1988). Positive electron microscopy requires the presence of complete viral assembly and positive immunohistochemistry requires the presence of viral capsid proteins. Viral DNA may be present within the host cells without these features although such manifestations of late gene expression are necessary for the virus to be infectious.

Although now more than 50 types have been described, (Smith & Campo, 1985; zur Hausen, 1987) in numerical terms five HPV types namely 6, 11, 16 and 18 (and to a lesser extent HPV type 31) are particularly important in the lower female genital tract. The majority of benign anogenital condylomata contain human papillomavirus type 6 and/or 11 whereas 70% or more of genital squamous cancers can be shown to contain types 16, 18 or 31. The preinvasive lesions of squamous epithelium (vulval, vaginal or cervical intraepithelial neoplasia) may exhibit a variety of viral types but those containing type 16 are more likely to progress (Campion et al., 1986). Current strategies are directed towards identifying women with cervical intraepithelial neoplasia associated with type 16, since these patients may be at a greater risk of malignant change and therefore require closer follow-up.

HPV types 6 and 11 are also implicated in the aetiology of laryngeal papillomatosis which will be discussed in more detail below.

Results

Human papillomavirus

In our studies of human papillomavirus by *in situ* hybridisation we have examined a wide range of anogenital lesions. Our most important original findings have included the detection of human papillomavirus type

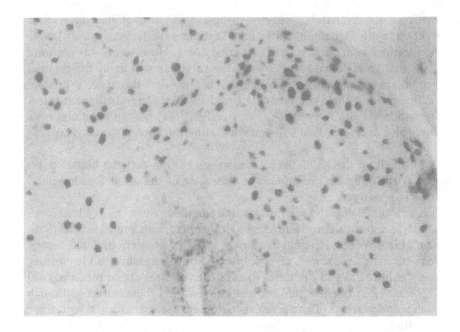

Figure 1. Positive *in situ* hybridisation signal for human papillomavirus type 11 in a case of juvenile multiple papillomatosis.

16 in squamous carcinomas of the anal canal in both male and females and the detection of HPV 6 and 11 in rare cases of malignant transformation of giant anogenital condylomata (Wells *et al.*, 1987, 1988).

We have also investigated the viral status of a large series of laryngeal papillomata (Terry *et al.*, 1987, 1989). Laryngeal papillomatosis is a distressing condition which may occur in childhood or in the adult. Multiple laryngoscopies or in rare cases even laryngectomy may be necessary to control the disease which may be recurrent over many years. Malignant change to laryngeal squamous carcinoma, however, appears to be a very rare phenomenon. In our most recently completed study of laryngeal papillomata, 8 of 14 patients with juvenile onset laryngeal papillomata, and 20 of 31 patients with adult onset laryngeal papillomata, have shown the presence of human papillomavirus types 6 and/or 11 (Fig. 1). Over 100 tissue blocks were examined and several specimens from individual patients accrued over many years were subject to *in situ* hybridization. The important finding from this extensive study is that patients with multiple confluent lesions were more likely to exhibit a positive hybridisation signal for HPV types 6 and/or 11, and those whose histology showed florid koilocytosis, and who showed strongly positive biopsy specimens, were more likely to have uncontrolled disease with a poor prognosis requiring multiple endoscopies. Furthermore, the virus seems to persist within laryngeal squamous epithelium over several years (Quiney et al., 1989).

The specific hybridisation signal is seen focally in the nuclei of superficial cells and correlates well with koilocytosis. Human papillomavirus types 6 and 11 show homology in their DNA sequences in the order of 85% and it seems that there is some cross-reactivity for these types on in situ hybridisation. However, we have never encountered cross-reactivity between 6/11 and 16 or between 16 and 18.

It is probably the basal cells of squamous epithelium that first become infected by the human papillomavirus. Initially, circular viral DNA is present in an episomal or extrachromosal form within the nucleus. It seems likely that the diffuse nuclear signal usually obtained with this technique is indicative of episomal DNA. Subsequent integration of viral DNA within host DNA in a linear fashion with subsequent transcription of viral sequences seems to be an important step in malignant transformation. More recently our attention has also turned to laryngeal carcinoma. Ten out of 29 (35%) cases were positive for HPV. Six cases were positive for HPV 18 only, two were positive for HPV type 31 only and two cases were positive for HPV types 18 and 31 (unpublished data). In many cases and particularly within the tumour itself a particulate signal consisting of several discrete dots was observed (Fig. 2). It is tempting to suggest that this type of reactivity reflects integrated viral DNA and a number of groups are claiming this.

Cytomegalovirus, JC virus and Epstein–Barr virus

The technique developed with human papillomavirus is working successfully and presently we are using a wide range of other probes, for example, a Y chromosome probe for the sexing of chorionic villus biopsy specimens (Thornton et al., 1989) and spermatozoa and probes to Cytomegalovirus (CMV), the JC virus and Epstein-Barr virus. We have used probes to Cytomegalovirus to demonstrate viral DNA in cases of intrauterine fetal death due to extensive fetal and placental CMV infection (Fig. 3).

JC virus can be demonstrated in the brains of patients with progressive multifocal leucoencephalopathy by in situ hybridisation (Ironside et al., 1989). This is a disorder which usually affects immunocompromised hosts (e.g. patients with lymphoma, sarcoidosis, AIDS or patients on immunosuppressive therapy). The condition is due to reactivation of latent virus which results in a productive (lytic) infection of oligodendrocytes with resulting demyelination (Fig. 4).

Recently we have completed an in situ hybridisation study of Epstein–Barr virus in 24 examples of primary central nervous system lymphoma (Murphy et al., 1990). Using three probes to various regions of the Epstein–Barr virus

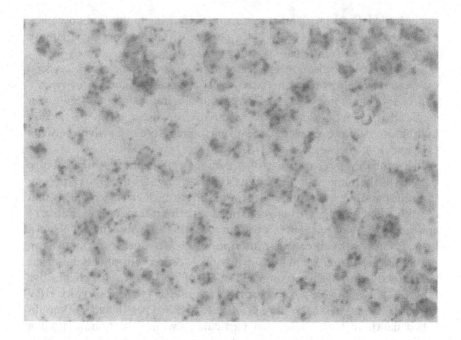

Figure 2. 'Particulate' or 'multiple dot' pattern of hybridisation signal for human papillomavirus type 18 in a laryngeal squamous carcinoma.

genome, a 'particulate; or 'multiple dot' pattern of hybridisation was present in the tumour cells of 11 cases while the surrounding brain cells were negative. These results suggest that the Epstein–Barr virus may be implicated in this particular type of lymphoproliferation.

Methodology

Our initial studies involved the development and application of an alkaline phosphatase detection system.

There are, however, several disadvantages associated with such a system:

1. The incubation times of up to 4 hours in the substrate that are sometimes necessary to achieve high sensitivity can give unacceptable levels of background reactivity.
2. The coloured reaction product is unstable when stained sections are dehydrated, cleared and mounted in synthetic mounting media. Sections are best mounted in aqueous mountants.
3. Some counterstains tend to leach out into the aqueous mountant, thus the range of counterstains is limited.

Demonstration of viruses by *in situ* hybridisation

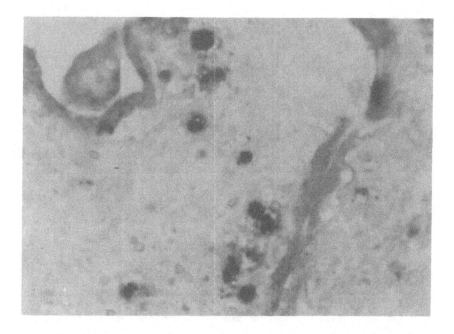

Figure 3. Positive *in situ* hybridisation signal for Cytomegalovirus in placental chorionic villi in a case of intrauterine fetal death.

In an attempt to eliminate these disadvantages we have devised an immunogold silver staining technique (Holgate *et al.*, 1983), based on the method published by Lewis *et al.*, (1987) (Jackson, Lewis & Wells, 1989). This technique has given reproducible and consistent results with the Y probe in the Raji cell line and with HPV probes in formalin-fixed paraffin embedded sections of human juvenile laryngeal papillomas.

The immunogold silver staining method offers the following advantages over the more usual detection methods.

1. It produces an intense black reaction which is insoluble in all commonly used dehydrating and clearing reagents.
2. It allows the use of a greater number of counterstains.
3. It produces a signal that can be enhanced by epipolarisation.
4. There are no known health hazards associated with the reagents used.

All the probes used in this study were nick translated with biotin 11 deoxyuridine triphosphate. The use of a streptavidin gold complex has also been investigated. This method, in our hands, has failed to give any demon-

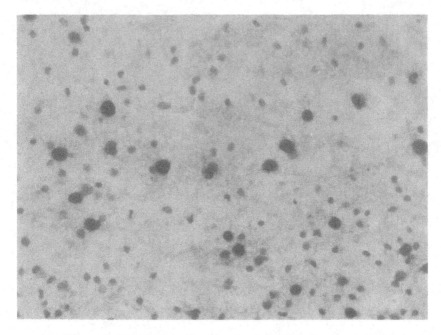

Figure 4. Positive hybridisation signal for JC virus in astrocytes and oliogodendrocytes in a case of progressive multifocal leucoencephalopathy (courtesy of Dr J. W. Ironside).

stration with the biotinylated probes used in the study. This may be explained by the large molecular size of the streptavidin gold complex. Stearic hindrance may make the binding of this complex impossible when the biotin is positioned on only a short 11 atom spacer arm. Nick translation of probes with a biotin nucleotide incorporating a longer spacer arm may lead to the successful application of a streptavidin-gold detection system.

There are a number of other ways in which the *in situ* hybridisation method might be improved.

1. The sensitivity of the polyalkaline phosphatase detection system can be increased as can any conventional immunohistochemical reaction by increasing the layers in the histochemical sandwich (Pringle *et al.*, 1987).
2. The process of hybridisation may be expedited by the use of microwave technology (Coates *et al.*, 1987).
3. It is also possible to use bromodeoxyuridine labelled probes that are subsequently detected by a monoclonal antibody to bromodeoxyuridine (Niedobitek *et al.*, 1988). Bromodeoxyuridine is a thymidine analogue and its use overcomes the problems that

may occur with biotinylated probes due to the presence of endogenous avidin-binding activity in frozen sections of some tissues.
4. mRNA can also be detected in tissue sections by *in situ* hybridisation though there are considerable technical difficulties because mRNA is rapidly denatured following removal of tissue from the host (Burns *et al.*, 1987; Stoler & Broker 1986; McDougall, Myerson & Beckmann, 1986).
5. One must concede that *in situ* DNA hybridisation may not always be as sensitive as Southern blot hybridisation and we may therefore be limited in our detection of viral or other genes when they are present in cells in low copy number. One way in which to overcome this problem is to amplify the DNA of the sequence that is being investigated. Currently there is growing interest in the POLYMERASE CHAIN REACTION a technique developed by geneticists at the Cetus Corporation. In this reaction sequences of DNA are amplified over a million times in a few hours. The principle of this method is described in Chapter 1 of this volume: the adaptation of this technique to *in situ* hybridisation will obviously have enormous application to diagnosis and pathology (see also Herrington *et al.*, this volume).

Protocols

In situ hybridisation with an alkaline phosphatase detection system

Human papillomavirus DNA probes in plasmid vectors were obtained as a gift from Professor Harald zur Hausen in Heidelburg. These plasmids are transfected into *Escherichia coli*, grown up, harvested and purified. The purified probe must then be labelled by nick translation. Briefly the probes are biotinylated with biotin-II-dUTP (Gibco-BRL) by using a nick translation kit (Gibco-BRL) and following the recommended protocol. Unincorporated dNTPs are separated from the biotinylated DNA by the spun column technique on Sephadex G50 (Pharmacia Ltd, Milton Keynes).

Pretreatment

Sections 3–5 μm thick from routinely formalin fixed and processed tissue are placed on 3-aminopropyltriethoxysilane coated single well slides and warmed up on a hotplate for at least one day to ensure maximum tissue adhesion. Sections are dewaxed in xylene at 37 °C for 30 min, xylene at room temperature for 10 min and absolute alcohol for 2 × 10 min. The sections are then hydrated through a series of graded alcohols to distilled water and immersed in phosphate buffered saline for 5 minutes. The slides are then transferred to 0.2 M HCl for 10 minutes, incubated in 0.1% Triton X-100 for three minutes to permeabilise membranes and digested with the protease Proteinase K (Gibco, Paisley, Scotland), for 15 min at 37°C to access the cellular DNA. After washing in several changes of phosphate buffered saline the slides are placed in 20% acetic acid in water at 4°C for 15 s to destroy endogenous tissue alkaline phosphatase and post fixed in 4% paraformaldehyde prior to dehydration through graded alcohols to 100% ethanol.

Hybridisation

The biotinylated probes are prepared at a concentration of 200 ng/ml in a hybridisation buffer containing $2 \times$ SSC, 5% dextran sulphate, 0.2% dried milk powder, 50% formamide ($1 \times$ SSC = 0.15 M sodium chloride, 0.015 M sodium citrate). 75µl hybridisation mixture (i.e. 12 ng DNA probe) are added to each prepared section, the well is covered with a piece of Gel Bond (ICN Biochemicals, High Wycombe, Bucks) hydrophobic side down, and sealed with nail varnish. Cellular and probe DNA are made single stranded by heating the slides at 90°C for 10 mins, and hybridisation is carried out at 42°C for 18–24 h in a humid box.

Detection of hybridisation signal

After carefully removing the Gel Bond cover slip, slides are immersed in 3% bovine serum albumin in Buffer 1 for 5 min to block all non-specific streptavidin binding sites in the tissue. Buffer 1 is composed of 0.1 M Tris-HCl pH 7.5, 0.1 M NaCl, 2 mM $MgCl_2$ and 0.5% Triton X-100. Streptavidin (2 µg/ml) in buffer 1 is then added to each slide and incubated for 20 min. This protein tetramer binds strongly with biotin. The slides are washed free of excess streptavidin and incubated with biotinylated alkaline phosphatase in Buffer 1 at a concentration of 1 µg/ml for 20 min to react with the remaining biotin binding sites on the bound streptavidin. The slides are washed in Buffer 1. Sections are then equilibrated in Buffer 2 for 30 min. Buffer 2 is composed of 0.1 M TRIS-HCl at pH 9.5, 0.1 M NaCl and 50 mM $MgCl_2$. The alkaline phosphatase is detected by reaction with a substrate containing nitroblue tetrazolium and 5-bromo-4-chloro-3-indolyl phosphate to produce a purple blue precipitate at the site of hybridisation. The slides are then counterstained with methyl green.

Immunogold–silver staining

Following *in situ* hybridisation the gel bond covering the wells is carefully removed and the slides immersed in $2 \times$ SSC. Sequentially the slides are washed in $2 \times$ SSC at 60°C for 20 min, $0.2 \times$ SSC at 42°C (moderate stringency) for 20 min and $0.1 \times$ SSC for 10 min prior to detection of the hybridisation signal.

Detection of hybridisation signal
1. Hybridised sections are placed in Lugol's iodine for 2 min, rinsed in water and decolourised in 2.5% aqueous sodium thiosulphate. Sections are then washed in tap water followed by Tris buffered saline pH 7.6 for 5 min.
2. Sections are treated with rabbit anti biotin (Enzo Biochem Inc., New York, USA) diluted 1 in 20 in Tris buffered saline (pH 7.6) for 60 min.
3. Sections are washed in Tris buffered saline (pH 7.6) for 5 min.
4. Sections are treated with normal goat serum for 5 min.
5. Excess normal goat serum is shaken off and goat anti-rabbit G5LM grade immunogold reagent (Janssen Pharmaceuticals, Beerse, Belgium) diluted 1 in 120 in Tris buffered saline pH 8.2 for 2 h, added.
6. Sections are washed in Tris buffered saline (pH 7.6) for 2×5 min.
7. Sections are washed in distilled water for 2×5 min.
8. Further silver enhancement of the gold labelled antibody is obtained by using a physical developing solution (Holgate *et al.*, 1986) or Intense II (Jansson Pharmaceuticals, Beerse, Belgium). Silver intensification is carried out until sections appear optimally developed under light microscopical control.
9. Sections are washed in distilled water for 5 min.
10. Sections are fixed in 2.5% sodium thiosulphate for 3 min.
11. Sections are washed in tap water for 1 min.
12. Counterstain as desired.
13. Sections are dehydrated, cleared and mounted in synthetic resin.

References

Burns, J., Graham, A. K., Frank, C., Fleming, K. A., Evans, M. F. & McGee, J. O'D. (1987). Detection of low copy human papilloma virus DNA and mRNA in routine paraffin sections of cervix by non-isotopic *in situ* hybridisation. *Journal of Clinical Pathology*, **40**, 858–64.

Campion, M. J., McCance, D. J., Cuzick, J. & Singer, A. (1986). Progressive potential of mild cervical atypia: prospective cytological, colposcopic and virological study. *Lancet*, **ii**, 237–40.

Coates, P. J., Hall, P. A., Butler, M. G. & D'Ardenne, A. J. (1987). Rapid technique of DNA-DNA *in situ* hybridization on formalin fixed tissue sections using microwave irradiation. *Journal of Clinical Pathology*, **40**, 865–9.

Crum, C. P., Nuova, G., Friedman, D. & Silverstein, S. J. (1988). A comparison of biotin and isotope-labelled ribonucleic acid in genital precancers. *Laboratory Investigation*, **58**, 354–9.

Gallyas, F., Görcs, T. & Merchenthaler, I. (1982). High grade intensification of the end product of the diaminobenzidine reaction for peroxidase histochemistry. *Journal of Histochemistry and Cytochemistry*, **30**, 183–4.

Gupta, J., Gendelman, H. E., Nagashfar, Z., Gupta, P., Rosenshein, N., Sawada, E., Woodruff, J. D. & Shah, K. (1985). Specific identification of human papillomavirus type in cervical smears and paraffin sections by *in situ* hybridization with radioactive probes: a preliminary communication. *International Journal of Gynecological Pathology*, **4**, 211–18.

Holgate, C. S., Jackson, P., Cowen, P. N. & Bird, C. C. (1983). Immunogold silver staining: new method of immunostaining with enhanced sensitivity. *Journal of Histochemistry and Cytochemistry*, **31**, 938–44.

Holgate, C. S., Jackson, P., Pollard, K., Lunny, D. & Bird, C. C. (1986). Effect of fixation on T and B lymphocyte surface membrane antigen demonstration in paraffin processed tissue. *Journal of Pathology*, **149**, 293–300.

Ironside, J. W., Lewis, F. A., Blythe, D. & Wakefield, E. A. (1989). The identification of cells containing JC papovavirus DNA in progressive multifocal leukoencephalopathy by combined *in situ* hybridisation and immunocytochemistry. *Journal of Pathology*, **157**, 291–7.

Jackson, P., Lewis, F. A. & Wells, M. (1989). *In situ* hybridisation technique using an immunogold silver staining system. *Histochemical Journal*, **21**, 425–8.

Langer, P. R., Waldrop, A. A., Ward, D. C. (1981). Enzymatic synthesis of biotin labelled polynucleotides: novel nucleic acid affinity probes. *Proceedings of the National Academy of Sciences*, USA, **78**, 6633–7.

Lewis, F. A., Griffiths, S., Dunnicliffe, R., Wells, M., Dudding, N. & Bird, C. C. (1987). Sensitive *in situ* hybridisation technique using biotin–streptavidin–polyalkaline phosphatase complex. *Journal of Clinical Pathology*, **40**, 163–6.

McAllister, H. A. & Rock, D. L. (1985). Comparative usefulness of tissue fixatives for *in situ* viral nucleic acid hybridization. *Journal of Histochemistry and Cytochemistry*, **30**, 1026–32.

McDougall, J. K., Myerson, D. & Beckmann, A. M. (1986). Detective of viral DNA and RNA by *in situ* hybridization. *Journal of Histochemistry and Cytochemistry*, **34**, 33–8.

Murphy, J. K., Young, L. S., Bevan, I., Lewis, R. A., Ironside, J. W., O'Brien, C. J. & Wells, M. (1990). Demonstration of Epstein–Barr virus in primary brain lymphoma by *in situ* DNA hybridisation in paraffin embedded tissue. *Journal of Clinical Pathology* (in Press).

Nagai, N., Nuovo, G., Friedman, D. & Crum, C. P. (1987). Detection of papillomavirus nucleic acids in genital precancers with the *in situ* hybridization technique. *International Journal of Gynecological Pathology*, **6**, 366–79.

Niedobitek, G., Finn, T., Herbst, H., Bornhoft, G., Gerdes, J. & Stein. H. (1988). Detection of viral DNA by *in situ* hybridization using bromodeoxyuridine labelled DNA probes. *American Journal of Pathology*, **131**, 1–4.

Pringle, J. H., Homer, C. E., Warford, A., Kendall, C. H. & Lauder, I. (1987). In situ hybridization; alkaline phosphatase visualization of biotinylated probes in cryostat and paraffin sections. *Histochemical Journal*, **19**, 488–96.

Quiney, R. E., Wells, M., Lewis, F. A., Terry, R. M., Michaels, L. & Croft, C. B. (1989). Laryngeal papillomatosis; correlation between severity of disease and presence of HPV 6 and 11 detected by *in situ* DNA hybridisation. *Journal of Clinical Pathology*, **42**, 694–8.

Sato, S., Okagaki, T., Clark, B. A., Twiggs, L. B., Fukushima, M., Ostrow, R. S. & Faras, A. J. (1987). Sensitivity of koilocytosis, immunocytochemistry, and electron microscopy as compared to DNA hybridization in detecting human papillomavirus in cervical and vaginal condyloma and intraepithelial neoplasia. *International Journal of Gynecological Pathology*, **5**, 297–307.

Singer, R. H. & Ward, D. (1982). Actin gene expression visualised in chicken muscle tissue culture by using *in situ* hybridisation with biotinylated nucleotide analog. *Proceedings of the National Academy of Sciences, USA*, **79**, 7331–5.

Smith, K. T. & Campo, M. S. (1985). The biology of papillomaviruses and their role in oncogenesis. *Anticancer Research*, **5**, 31–48.

Stoler, M. H. & Broker, M. H. (1986). *In situ* hybridization detection of human papillomavirus DNAs and messenger RNAs in genital condylomas and a cervical carcinoma. *Human Pathology*, **17**, 1250–8.

Terry, R. M., Lewis, F. A., Griffiths, S., Wells, M. & Bird, C. C. (1987). Demonstration of human papillomavirus types 6 and 11 in juvenile laryngeal papillomatosis by *in situ* DNA hybridization. *Journal of Pathology*, **153**, 245–8.

Terry, R. M., Lewis, F. A., Robertson, S., Blythe, D. & Wells, M. (1989). Juvenile and adult laryngeal papillomata: classification by *in situ* hybridisation for human papillomavirus. *Clinical Otolaryngology*, **157**, 109–15.

Thornton, J. G., Lewis, F. A., Linton, G., Wells, M., Tyrrell, S. & Lilford, R. J. (1989). Fetal sexing by chorionic villus biopsy and *in situ* DNA hybridization with a Y probe and biotin-streptavidin polyalkaline phosphatase labelling. *Journal of Obstetrics and Gynaecology*, **10**, 1–4.

Wells, M. Griffiths, S., Lewis, F. A. & Bird, C. C. (1987). Demonstration of human papillomavirus types in paraffin processed tissue from human ano-genital lesions by *in situ* DNA hybridization. *Journal of Pathology*, **152**, 77–82.

Wells, M., Robertson, S., Lewis, F. & Dixon, M. F. (1988). Squamous carcinoma arising in a giant peri-anal condyloma associated with human papillomavirus types 6 and 11. *Histopathology*, **12**, 319–23.

Wilbur, D. C., Reichman, R. & Stoler, M. H. (1988). Detection of infection by human papillomavirus in genital condylomata. A comparison study using immunocytochemistry and *in situ* nucleic acid hybridization. *American Journal of Clinical Pathology*, **89**, 505–10.

Zur Hausen, H. (1987). Papillomaviruses in human cancer. *Cancer*, **59**, 1692–6.

INDEX

AAF 34
abd-A 100
accessibility 34
acetic anhydride 15
actin CyI 81
actin CyIIa 81
algae 143
alkaline phosphatase 11, 13, 185, 225, 279, 280
allyl-UTP 9
alphoid repeats 245, 258
AMCA 36
aminoacetylfluorene 10, 242
amniotic fluid cells 209
amoeba 144
Anl 116
aneuploidy 258
ANP 56, 60
antenatal sex determination 259
Antennapedia 100
anti-biotin antibodies 11
anti-fading agents 35
atrial natriuretic peptide (ANP) 43, 44, 46, 56
atrium 56
autofluorescence 36
autoradiography 48, 62, 127, 139, 209, 228, 230
avidin 34, 184

B-tubulin 81
bacterial detection 2, 22
bcd 103
bicod (bcd) gene 103
biopsy material 241
biotin 224
biotin 11-dUTP 10, 34, 279
biotin–avidin 10
biotinylated probes 177
bithorax 100
blastocysts 214
blastomeres 71

blastula 71
Bluescript 99
brain 53, 55, 56
bromodeoxyuridine 10, 12

calcitonin gene-related peptida (CGRP) 43, 44, 46, 49, 60
caudal (cad) 108
cDNA 157
cDNA libraries 138
cell biology 37
cell spreads 233
central nervous system (CNS) 134
Chlorarachnion 144
chloroplasts 143
chorionic villus cells 209
chromosomal aberrations 22
chromosome 241, 245, 257, 259
chromosome abnormalities 259, 260
chromosome mapping 22
chromosome painting 35
chromosome specific libraries 35, 258
chromosome specific probes 257, 258
CMV 249
CNS 135
colloidal gold 13, 35, 155
comparison of methods 180
competitor DNA 35
confocal laser scanning microscopy 37
controls 17, 49, 244
cosmid clones 35
cryptonomads 143
cyanins 36
CyIIIa 74
cytogenetics 37
cytomegalovirus (CMV) 37, 275, 248
cytopathic effect 251

DAPI 13
denaturation 35
detection systems 244
detergent treatments 34

286 Index

dextran sulphate 16
diagnosis of viral disease 251
differential screening 159
diffuse neuroendocrine system 44, 43, 58
digoxigenin 34, 99, 224
dinitrophenol 13
direct methods 33
DNA 1, 247
DNA polymerase 6
DNA–DNA 3
DNA–RNA 3, 13
DNA–RNA hybrid antibodies 10
DNAse 1 6
DNAses 18
dorsal root ganglia 49, 51, 53
dot blot 158, 271
double haptenised probes 36
double labelling NISH 256
double stranded DNA 2
double-labelled probe 231
Drosophila 99
duplexes 3

ectoderm 79, 80
electron microscopy 153, 254
embryonic lethal mutations 103
embryos 212
endogenous biotin 184
endosperm 178
endosymbiosis 143
enteric neurones 51, 53
enzymes 35
Epstein–Barr Virus (EBV) 249, 275
even skipped (eve) 105

false hybridisation 162
fetal lymphocytes 209
fetus 208
filter hybridisation 34
filter *in situ* hybridisation 256
fine mapping 35
FITC 36
fixation 14, 21, 61, 69, 123, 139, 169
flow cytometry 24
flowering 157, 225
fluorescence microscopy 35, 226
fluorescence ratio imaging 36
fluorescein 13
fluorochrome 10, 35
formamide 3, 16
freeze–thaw 34
functional significance 107
fushi tarazu (ftz) 105

gap genes 103, 105
gastrulation 74

GC content 3
gene amplification 37
gene mapping 2
genetics 131
Giemsa banding 37
glutaraldehyde 69
gradient 108

hapten 33
hapten modifications 33
HboxI 86
Hepatitis B virus 249
herpes virus 248
Hg 34
homoeotic mutations 100
horseradish peroxidase 224
human papillomavirus (HPV)6/11/16/18/
 31/33/35 37, 248, 256, 257, 271, 273
hunchback (hb) 105
hybrid stability 2
hybridisation 21, 48, 62, 126, 139, 280
hybridisation kinetics 2
hybridisation of whole tissues 99
hybridisation times 3
hydrogen bonds 2

immunocytochemistry 178, 241
immunoglobulins 34
Immunogold–silver 280
immunohistochemistry 17, 254
immunological amplification procedures 35
immunological response 251
in situ suppression hybridisation 258, 259
indirect techniques 33
int-1 132, 133, 134, 135
interphase cytogenetics 245, 257, 259, 260, 261
interphase nuclei 37
ionic strength 3

JC virus 37, 249, 275

Klenow polymerase 7
knirps (kni) 105
Krox-20 132, 134, 138
Kruppel (Kr) 105

legumin 178
limitations 97
LINEs 245, 258
localisation 115
LR Gold resin 153
Lymnaea stagnalis 38

M13 bacteriophage vectors 8
maize 157

Index

male flower specific genes 157
marker chromosomes 261
maternal effect mutations 103
matertnal mRNAs 115
meiosis 158
Mercurated nucleotide 10, 12, 242
mesenchyme 72
Micrococcal nuclease 19
microspores 165
midisatellite sequences 245
midisatellites 259
miniature nucleus 144
mounting sections 125
mouse embryogenesis 131
mRNA 2, 250, 251
mRNA localisation 22, 37
multiple hybridisation 33
multiple oligonucleotide 18
myocytes 56, 60

N-acetoxy-2-acetylaminofluorene 13
neuromeres 135
neuronal specification 110
neuropeptide Y (NPY) 43, 46, 55
neuropeptidergic systems 38
nick translation 6, 242
non-isotopic methods 1, 10, 17, 99, 241, 257
NPY 55, 56
nuclear proteins 74, 78
nuclear RNA 249
nucleus 143
numerical apertures 36

oligo-riboprobes 24
oligonucleotide primer extension synthesis 7
oligonucleotides 2, 8, 242
oncogene detection 22
oncogenes 2
oncology 37
oocytes 115
oogenesis 116

pair rule 103
pair rule genes 105
Papovavirus 248
parvo virus B19 37
pepsin 15, 34
peptides 43, 45, 46, 55, 58, 60
peptide mRNAs 46
permeabilisation 15
peroxidase 13
pGem 99
Photobiotin 7
pituitary 56, 58, 60

plastid 143
pole cells 80
Polymerase Chain Reaction (PCR) 4, 216
polyploid nuclei 219
polyU 69
post hybridisation washes 3, 16, 62, 48
post-embedding 184
pre-embedding 184
pre-embryo biopsy 217
pre-embryos 212
pre-hybridisation 21, 48, 62, 125
prenatal determination of sex 22, 260
prenatal diagnosis 205, 208
preparation of small cRNA probes by limited alkaline hydrolysis 61
pretreatment 139, 279
probe labelling 5, 242
 ^{32}P 126, 241
 AAF 34
 allyl-UTP 9
 aminoacetylfluorene 10, 242
 biotin 224
 biotin 11-dUTP 10, 34, 279
 bromodeoxyuridine 10
 digoxigenin 224
 double-labelled probe 231
 Hg 34
 mercurated nucleotide 10
 Photobiotin 7
 sulphonated nucleotide 10
probe production 4, 46
probe synthesis 126
prolactin 43, 46, 56, 58, 60
Pronase 15
pronuclei 71
Proteinase K 15, 34, 70

quality control 34

random priming 242
redundant transcriptional activity 108
Reflection Contrast Microscopy 13
repetitive sequences 37
reporter molecules 13, 33
restriction fragments 1
rhodamine 13
riboprobe 46, 47, 99
ribosomes 144
RNA 1
RNA (cRNA) probes 8, 43, 44, 45, 70
RNA polymerases 9
RNA viruses 251
RNA–RNA 3
RNAse 17, 18, 34
rRNA 143

288   Index

salmon sperm DNA 16
sea urchin 69
secondary controls 20
section loss 15
segment polarity 103
segmentation gene 104
sensitivity 37, 246, 247, 261
sex 2
Sex combs reduced (Scr) 101
sex-linked disorders 206
sex mismatched transplants 259, 260
silver enhancement 13, 185, 281
SINEs 245, 258
single copy genes 241, 245
single copy gene sequences 37
single stranded DNA 2
S1 nuclease 17
somatic cell hybrids 37
SP6 9, 70
SpARS 85
Spec1 71, 74
Spec3 81
specific activity 6
spermatozoa 219
spinal cord 49, 51, 53
SSC 16
steric hindrance 3
storage-protein 175
streptavidin 10, 11
streptavidin gold 185
stringency 243, 256
structural chromosome aberrations 37
substance P 43, 44, 46, 51
sulphonated nucleotide 10
sulphonation/transamination 12, 34
synthesis of single-stranded cRNA probes 60
synthetic polynucleotide kinase oligonucleotides 35

T3 9
T4 8

T7 9
target sequence 34
terminal deoxynucleotide transferase 8
terminal transferase 34
tetraploid nuclei 209
Texas Red 13, 36
time-resolved luminiscent reporter molecules 36
tissue preparation and processing 46
tissue pretreatments 14, 139, 179
Tm 2, 16
topological resolution 33
transfected and transgenic DNA 37
translocation 258
TRITC 36
trophectoderm 219
trophoblast cells 223

ultrastructural 181
unique DNA sequence 36
unmasking 243

vasoactive intestinal polypeptide (VIP) 43, 44, 46, 53
very early blastula (VEB) 76
Vgl 111, 116
viral DNA 247, 271
viral culture 254
viral detection 1, 22
virology 37
VNTR(s) 258, 259

wax embedding 162

Xenopus 115
XRITC 36

Y probe 208
yeast artificial chromosome (YAC) 35

Printed in the United States
By Bookmasters